AF703238

Ontoform and Intrinsics of Subelemental and Chemospheric Superstructons

Eirenaios G

Ontoform and Intrinsics of Subelemental and Chemospheric Superstructons

First Edition October 2025

Written by Eirenaios

Contents

Chapter 3

Chapter 4

Chapter 7

Chapter 1

Introduction, Methodology and Scope of the Thesis

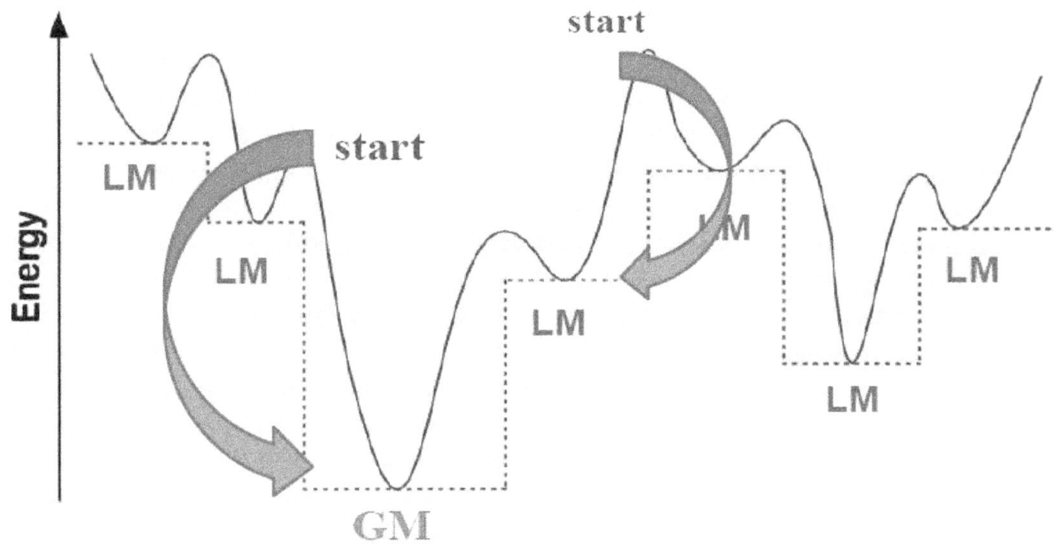

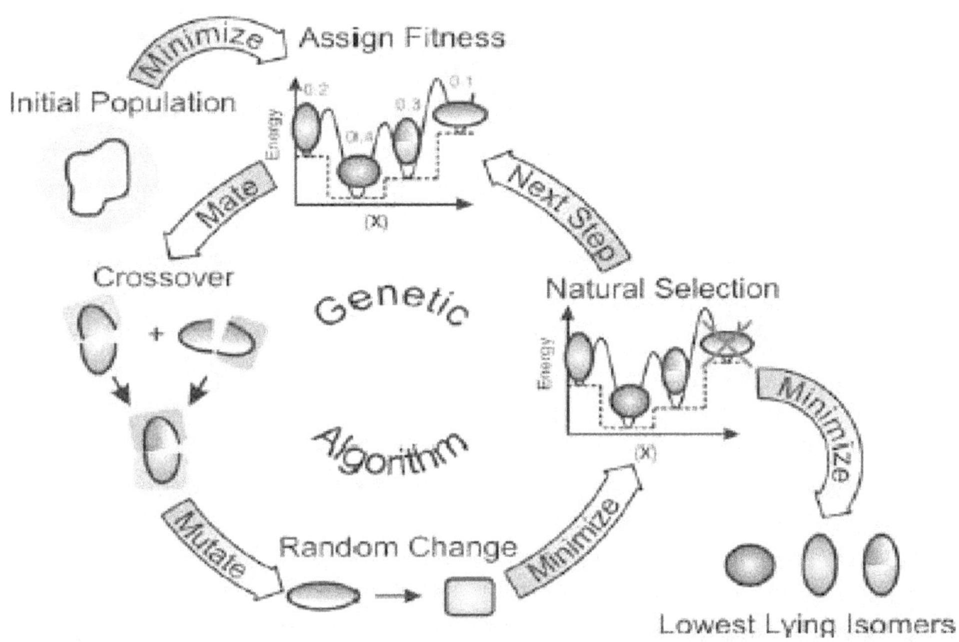

1.1 Introduction

Clusters are aggregates of between a few and thousands of atoms or molecules. Interest in cluster arises, because they constitute a new type of material which may have properties which are distinct from those of discrete molecules or bulk matter. Another reason for the interest in clusters is the size-dependent evolution of cluster properties [1-3]. Research on cluster is developing rapidly, which is an interdisciplinary of atomic physics, quantum chemistry, and many other disciplines. The atomic and molecular clusters have attracted numerous studies for their possible applications in catalysis, medicine, and nanoelectronics [4-8]. The clusters constitute a new type of material whose physical and chemical properties are distinct from those of discrete molecules or bulk matter. The size dependent evolution of geometric and electronic structures of clusters is also one of the reasons for the intensive interest. Theoretical researches are of great importance in cluster science because existing experimental methods for structural determination can seldom obtain the structure of clusters directly and understanding of the spectroscopic and mass spectrometric results often needs appropriate theoretical models [9-15]. Since quantum chemical methods such as the density functional theory (DFT) and Hartree-Fock method are not competent for large-size (N>30) clusters due to the huge computational cost and numerous local optima,[9-15] developing efficient global optimization algorithms and empirical potential functions is crucial to predict and understand the structures and properties of clusters[13-17].

Structure determination is, perhaps, the most challenging contemporary problem in the field of nanoscience, where computational techniques are the only available tool. At the nanoscale even simple binary systems that are well characterised in the bulk state exhibit surprising diversity. To understand this behaviour we apply a combination of global optimization technique with quantum chemical method (Density Functional Theory), using a state-of-the-art treatment of exchange and correlation, to elucidate the energy landscapes underlying this phenomenon. Our study also illustrates important technical points relating to the problem of searching the configurational space of isolated clusters for the lowest, or most stable, atomic and molecular structures. The magnitude of this problem is very much dependent upon the number of atoms that make up the cluster, as well as how trial configurations are generated and how expensive it is to model their interactions. Typically, low lying local minima on the energy landscape defined by interatomic potentials (IPs) are found by employing a number of global optimisation algorithms [18,19]. The use of IPs, as opposed to density functional theory (DFT) reduces the computational cost of locating plausible structures which can be refined at a later stage.

There are two basic issues that need to be considered when developing candidate structures for a certain cluster. The first concerns the complexity of the overall optimization problem and the second is how to model the interatomic bonding adequately in a computationally efficient way. In this case the general problem is to explore the (3N+1) dimensional potential energy surface (PES) of the N-atom cluster and find several minima with the ultimate goal of locating the overall lowest lying stationary point in energy, defined as the global minimum (GM). Hence the task is to perform a global optimization (GO) of the potential energy as a function of all atom coordinates. Now, how one can achieve this goal? Of course, the simplest way is to use heuristic information about the bonding in the system under study, that is, to perform a biased structure search. For example, bonding motifs found in the solid state or in molecules could be used as possible starting structures. Unfortunately, the novel properties of nanomaterials are closely related to their unique bonding and, consequently, assuming structures based on known compounds could result in erroneous structure predictions. The larger the cluster gets the greater is the number of meta-stable isomers, and a rough estimate indicates that the number of stable structures rises exponentially with the number of atoms. Although this kind of approach can be used for smaller clusters, it becomes computationally infeasible for larger systems. The situation becomes even more complicated for so-called nanoalloys, which are clusters in which two or more metals are combined and whose properties can be tuned by varying their composition and chemical ordering, as well as their size and morphology. For these nanoalloy clusters, locating the GM is complicated by the existence of a large number of homotops (inequivalent permutational isomers) for each geometric isomer. Hence, the best approach is to perform an unbiased systematic exploration of the PES with a global optimization algorithm.

Besides the need for an efficient global optimization method, the other prerequisite to make reasonable predictions of cluster structures is to describe the PES realistically. Hence, the level of theory that is used for the actual global optimization has to accurately reflect the nature of the chemical bonding in the nano-sized aggregate. For large nanoparticles, it is expected that the bonding is closely related to that of the bulk. Consequently, for a structural exploration of these large systems, model or empirical potentials can be used, which approximately mimic the bonding in the solid state. The smaller the nanoparticle is, however, the more important a full quantum chemical description of the interatomic bonding becomes. Even for larger clusters, a quantum chemical description of the bonding may be necessary, for example, when electronic shell closure leads to enhanced stability for a particular cluster size. Hence the most general approach in theoretically predicting the structure of clusters and nanoparticles is to perform a global optimization investigation using empirical potentials followed by refinement of candidate low energy isomers using quantum chemical electronic structure methods. In this thesis , a brief overview of global optimization algorithm is given, though concentrating mainly on Simulated Annealing and Genetic Algorithm techniques, first in combination with empirical potential. The thesis deals with implementation and application of these search methods to allow exploration for global minimum structures directly using electronic structure methods and, in particular, density functional theory.

1.2 Geometry optimisation of clusters

Whether one is using empirical atomistic potentials, ab initio MO theory or Density Functional Theory (DFT) to describe the bonding in clusters, one of the prime objectives is to find, for a given cluster size, the arrangement of atoms(or molecules) corresponding to the lowest potential energy –i.e the global minimum (GM) on the potential energy hypersurface. Now 'why is it important to locate the global minimum?' Clusters corresponding to global minima (or low lying local minima) are the most likely candidates for the most probable structure formed in a cluster experiment [20]. Experiments alone often provide an incomplete picture of the cluster structure and only by combining them with independent theoretical investigations can provide a complete description of the geometric arrangement and the corresponding properties be established [21-23]. Now the problem is the larger the cluster gets, the greater is the number of isomers and the number of stable structures rises exponentially with the number of atoms. Hence the best approach is to perform an unbiased systematic exploration of the PES with a global optimization algorithm. Besides the need for an efficient global optimization method, the other prerequisite to make reasonable predictions of cluster structures is to describe the potential energy surface (PES) realistically. The strategy of using empirical PES, though elegant and computationally quick may not be that easy to use in practice. These surfaces support a large number of minima and the task of finding the lowest one from the myriad of possibilities is a nontrivial task. Conventional gradient-based optimization techniques are found to be inadequate in most cases as the structure which is generally found out is the one which is achieved by the quickest descent in gradient norm from the initial starting point. Thus, the guarantee that a global structure is reached is entirely fortuitous and one has to scan the entire search space starting from different initial guess geometries to arrive at the correct result. Hence there has been an increasing tendency to use truly global and stochastic optimization. In general, a vast number of global optimization methods exist that can all be combined with electronic structure calculations. The global optimization of high-dimension functions is often nondeterministic-polynomial-time (NP) hard. The number of local minima will grow exponentially with the number of atoms [13-16]. To conquer the difficulties of the large computational demands and numerous local optima brought by the high-dimension variables space, many global optimization schemes have been proposed in the passed decades. These algorithms can be categorized into two classes. The first class uses the iteration of single solution. Its representatives are basin hopping (BH) algorithm[24-31] and dynamical lattice searching (DLS) method.[16,17,32-35] BH method and its variants are the combination of Monte Carlo method and a deterministic local minimization procedure.[24–28] They are very simple with few control parameters and general to various optimization problems. DLS method consists of lattice construction by heuristic strategies, lattice searching by iteratively moving the atom with high energy to the lattice site with lower energy, and a local minimization procedure.[16,17,32–35] Compared to BH methods, the searching of DLS methods is a little faster but more heuristic.[16,17] Both the two algorithms and their modifications [27-29,33–35] have succeeded in finding the stable structures of many types of clusters.[24—35] The other class with Evolutionary Algorithms (EAs)[36-53] as the representative, adopts a population of candidate structures in the search. In EAs such as genetic algorithm (GA) [36–51], antcolony method [52] and particle swarm optimization [53] the iteration of the population mimics the biological evolution. Individuals cooperate with each other to search the global optima of the fitness function by crossover, mutation, recombination, and selection operators.[36] After introducing real-valued cartesian coordinates coding,[40,41] a gradient-driven local minimization[42– 44] and three-dimension ''cut-and-splice'' crossover operator,[42,43] EAs achieve remarkable successes in the optimization of clusters.[45–54]

The overall task of a GO is to find the GM on the PES of the nanoparticle system under study. Of course, it is far too time consuming to explore the complete PES with all its local minima and all connecting transition states (TS). To fulfill this task, several highly efficient GO algorithms have been developed. To locate the GM with MC or MD methods, it is important to increase and then decrease the temperature. Hence, these algorithms are equivalent to crystallizing the cluster structure slowly from a melt. For this so-called simulated annealing (SA) approach, MC and MD variants exist which can be applied to the global optimization of cluster structures. Theoreticians who work with model empirical potentials to study cluster systems face another stiff challenge. This is the challenge of getting deeper minima. These potential energy surfaces are extremely rugged and the search for good structures can be a problem for any deterministic optimization method. Hence there has been an increasing tendency to use truly global and stochastic optimization. Simulated Annealing (SA) and Genetic Algorithm (GA) are two such stochastic optimization methods. Simulated Annealing (SA) [55,56] borrows its concepts from metallurgical process of annealing. As in metallurgical annealing, there is a temperature like quantity which controls the thermal fluctuations while searching a rugged potential energy landscape. At high temperatures, the search, even if trapped in a local attractive basin, will be able to come out of it and surmount the barrier separating it from other minima and after having sampled the search space at different temperatures, will ultimately be able to unequivocally locate the global minimum in the limit of annealing temperature tending to zero. Genetic Algorithm (GA) is another stochastic search technique based on the principles of natural evolution. GA draws its working philosophy from the genetic processes of selection, crossover and mutation. A mathematical equivalent of selection, crossover and mutation have shown promise as a truly global and stochastic search algorithm, which is able to find out low energy structures of complex large molecules and clusters. Simulated Annealing and Genetic Algorithm are two stochastic optimization techniques which can be applied to global optimization of cluster structures. The details of Simulated Annealing (SA) and Genetic Algorithm (GA) methods are given in the next sections.

1.2.1 Simulated Annealing

Simulated Annealing [57-63] is a stochastic global optimizer whose operation is based on metallurgical process of annealing. In thermodynamic annealing, a molten mixture of metal at a very high temperature (known as annealing temperature) is created and then cooled gradually till the system reaches the thermodynamic minimum energy state. The simulation starts at a high temperature, T_{at} (annealing temperature) and then the annealing temperature is slowly decreased with a specific rate, known as annealing schedule. At high T_{at} more and more area of the search space is sampled and as the search proceeds gradually with the lowering of temperature, the sampling area is decreased and in the limit of $T_{at} \rightarrow 0$, the optimum solution is found out.

If one wants to find the global minimum of a potential energy function possessing multiple minima, one starts with a randomly generated guess which lies far away from the desired actual solution. For the present problem we ascribe cartesian co-ordinates to each atom and from it we get the values of r_{ij} between the ith and jth atom. So the actual variables to be optimized are the cartesian coordinates and in a system of n atoms a starting trial solution S_0 is given by

$$S_0 = x_1^0, x_2^0, x_3^0, \ldots\ldots x_n^0 \ldots\ldots y_1^0, y_2^0, y_3^0, \ldots\ldots y_n^0 \ldots\ldots z_1^0, z_2^0, z_3^0, \ldots\ldots z_n^0 \quad (1.1)$$

The aim of the optimizer would be to obtain the correct set of coordinates and if this occurs after N number of steps, then the correct solution is given by

$$S_N = x_1^N, x_2^N, x_3^N, \ldots\ldots x_n^N \ldots\ldots y_1^N, y_2^N, y_3^N, \ldots\ldots y_n^N \ldots\ldots z_1^N, z_2^N, z_3^N, \ldots\ldots z_n^N \quad (1.2)$$

The idea is to modify the starting solution following certain steps so that at the end of the search the global minimum solution is found out. The operational details of SA are as follows: From the initial chosen set (S_0), any one coordinate is selected randomly say x_k^0. This is subjected to a random change in the following manner:

$$x_k^1 = x_k^0 + (-1)^l \Delta \times R \quad (1.3)$$

Here l is a random integer, R is a random number between 0 and 1 and Δ is the amplitude of change of coordinate randomly in positive or negative direction. Once the new set of coordinates has been found out one can calculate U_1, which is the potential energy of this new set. This value is compared with the potential energy U_0 evaluated with older set of coordinates. Thus in a given step of optimization (ith step), the objective function is given by

$$F_i = (U_i - U_L)^2 \quad (1.4)$$

where U_L is a lower bound to the energy which is supplied, U_i is the potential energy in the i_{th} step of iteration and F_i is the magnitude of objective function. After random change of positional coordinates the objective function for $(i + 1)^{th}$ step is

$$F_{i+1} = (U_{i+1} - U_L)^2 \quad (1.5)$$

If ΔF (F_{i+1}-F_i) is negative, the move is accepted. If not the move is subjected to the metropolis test. In Metropolis test we calculate Metropolis sampling probability which is given by

$$P = \exp\left(\frac{-\Delta F}{k_b T_{at}}\right) \quad (1.6)$$

where T_{at} is a temperature like quantity known as Simulated Annealing temperature. The quantity P always lies between 0 and 1. Once 'P' is evaluated a random number R between 0 and 1 is invoked and if P is greater than the called random number the move is accepted. It is clear that if T_{at} is high, P will be close to 1 and more moves will pass the Metropolis test. The physical meaning is that at higher simulation temperature, due to strong thermal fluctuations, a greater length of search space is sampled and nearly all moves become accepted. This feature makes SA a global optimizer, since if one is trapped in a local attractive basin, then one has to temporarily increase the potential energy of the system before the search to the global can begin. For a given temperature T_{at}, the system is sampled a large number of times before decreasing the temperature and passing on to the next temperature step. This decrease of temperature is necessary as with the passage of number of steps the system gradually move towards the global minimum and large thermal fluctuations are not necessary. As the simulation proceeds, T_{at} is gradually decreased by following some schedule known as annealing schedule. The SA temperature can be thought of as a parameter which controls thermal fluctuations. If a search is trapped in a local attractive basin and needs to come out of it, the high T_{at} helps it to climb out of it, crossing the barrier separating it from the deeper minimum and allowing the system to move from one local attractive basin to another which is lower in energy. A high T_{at} means a higher value of metropolis sampling probability which helps the search to accept moves which involve rise of energy until a better path is found out taking the system to a deeper minimum. As the annealing temperature decreases during the latter stages of the search thermal fluctuations decrease, the Metropolis sampling probability becomes low and only those moves in which the energy decreases only are the ones that are accepted. This is the way the search should be at the end of the optimization process, since once the direction of global minimum is achieved, there is no need to surmount energy barrier and in the limit of $T_{at} \rightarrow 0$ the right desired solution is found out.

In the field of geometry optimisation there has been considerable amount of work in large systems involving simulated annealing. Simulated annealing has been effectively used in quantum chemical calculations, especially in the optimization of parameters of wave functions [64-66]. This method has also been used to solve problems such as to locate the critical points and to construct reaction paths for the transformations on a model potential energy surface and for rearrangement of Lennard Jones clusters.[67-68]. There has also been interesting report of using simulated annealing method [69] using interplanar angles and lattice spacings.

1.2.2 Genetic Algorithm

GA is a search algorithm based on the natural process of genetics and "Survival of the Fittest" as given by Darwin. GA uses analogues of selection, crossover and mutation operations to find the optimum solution. The simulation starts from a pool of trial solutions (generated randomly) and not a single one. The individual solutions are popularly known as strings. The simulation is started with solution pool containing strings of low fitness, which is defined below. These strings are iteratively subjected to the operations of selection, crossover and mutation, till a string is obtained with fitness value close to one (which is the desired solution). These randomly selected solutions will be far away from the minimum. How far the individual strings from the optimal solution are is described in terms of fitness function. The fitness of a particular ith string is given by

$$f_i = \exp\left\{-\left(U_{tot}^i - U_L\right)^2\right\} \tag{1.7}$$

where U_L is a lower bound to the energy and is supplied and U_{tot}^i is the energy of the ith string in the solution pool where the energy is calculated using the empirical potential energy function. To start with we create a pool 'M' number of randomly selected populations S_M.

$$S_1 = x_1^1, y_1^1, z_1^1, x_1^2, y_1^2, z_1^2, \dots\dots\dots x_1^k, y_1^k, z_1^k, \dots\dots\dots\dots x_1^N, y_1^N, z_1^N \tag{1.8}$$

$$S_2 = x_2^1, y_2^1, z_2^1, x_2^2, y_2^2, z_2^2, \dots\dots\dots x_2^k, y_2^k, z_2^k, \dots\dots\dots\dots x_2^N, y_2^N, z_2^N \tag{1.9}$$

.

.

$$S_M = x_M^1, y_M^1, z_M^1, x_M^2, y_M^2, z_M^2, \dots\dots\dots x_M^k, y_M^k, z_M^k, \dots\dots x_M^N, y_M^N, z_M^N \tag{1.10}$$

Where x_M^k, y_M^k, z_M^k are the x,y and z - coordinates of the k_{th} atom of the cluster in the Mth trial solution. Now obviously these randomly selected solutions will be far away from a minimum. As a first step to improve the set, we calculate the total energy for each solution, where U_{tot}^i is the general energy expression for a given i^{th} solution in the pool. Once this is done we calculate for each solution a quantity called the fitness function (f_i), where f_i is the ftness for a given i^{th} solution in the pool which is defined as

$$f_i = \exp\left\{-\left(U_{tot}^i - U_L\right)^2\right\} \tag{1.11}$$

where U_L is a lower bound to the energy and is supplied. Now 'f_i'can have values between 0 and 1 and for a solution whose U_{tot}^i is close to U_L, the fitness function will be close to 1. The lower the value of 'f_i' the further it is from the actual solution and vice-versa. While finding the global minimum, U_L is kept at a low enough value as possible, such that the fitness of the solutions in the pool will become large only when a really low energy minimum is obtained. One can tune the GA search to find higher energy structures also by updating U_L as U_L (new) $=U_L + \sigma$, where σ is a shift in the lower bound supplied from outside. With this new U_L {U_L(new)}, if a GA search is carried out; high energy structures will be found out since only a higher energy structure will correspond to a high fitness. Once the fitness function has been assigned for each solution, we do an operation called selection. The selection of better individuals from a set of 'M' trial solutions can be achieved in a number of ways. We have employed the so called Roulette Wheel method. In this a hypothetical wheel containing 'M' slots corresponding to each trial solution are imagined. However, the slots are not of equal width. Solutions with higher fitness have greater widths and vice-versa. Subsequently a random number is called 'n' times and the slot in which the random number falls is picked up for selection. Obviously the slot with larger width will have a higher probability of being picked up by this procedure than those with smaller width. As the width of the slot is dictated by the fitness value, large fitness solution will be picked up more and the weaker solution to a lesser degree. So after selection the average fitness of the solution pool will increase, but selection alone will not introduce new features into the starting pool, which is achieved by the use of other operators. To do this we employ two operators called crossover and mutation. In crossover a certain number of solutions (or strings as they are called in GA terminology) are selected and grouped pairwise. For a pair a location on the string called crossover site is selected randomly and all variables after that is exchanged with the other string in the pair. Schematically the process looks like this;

Before Crossover

$$S_J = x_J^1, y_J^1, z_J^1, x_J^2, y_J^2, z_J^2, \ldots\ldots\ldots\ldots x_J^l, y_J^l, z_J^l, \ldots\ldots\ldots\ldots x_J^N, y_J^N, z_J^N \qquad (1.12)$$

$$S_K = x_K^1, y_K^1, z_K^1, x_K^2, y_K^2, z_K^2, \ldots\ldots\ldots\ldots x_K^l, y_K^l, z_K^l, \ldots\ldots\ldots\ldots x_K^N, y_K^N, z_K^N \qquad (1.13)$$

If '*l*' is the crossover site then after crossover the two strings become;

After Crossover

$$S_J' = x_J^1, y_J^1, z_J^1, x_J^2, y_J^2, z_J^2, \ldots\ldots\ldots\ldots x_K^l, y_K^l, z_K^l, \ldots\ldots\ldots\ldots x_K^N, y_K^N, z_K^N \qquad (1.14)$$

$$S_K' = x_K^1, y_K^1, z_K^1, x_K^2, y_K^2, z_K^2, \ldots\ldots\ldots\ldots x_J^l, y_J^l, z_J^l, \ldots\ldots\ldots\ldots x_J^N, y_J^N, z_J^N \qquad (1.15)$$

Obviously now we have strings having particular sets of co-ordinates, which were absent in the originally selected trial solution set. Hence crossover can introduce new variety and information into the solution set. The other operator which can do a similar thing, albeit in a different way, is mutation. Mutation occurs infrequently and acts not on the string as a whole but on any individual variable, which in our case are the coordinates. If a variable is selected for mutation this is what the operation looks like;

Let x_M^i be the coordinate which has been identified for mutation. After mutation the variable $x_M^{i'}$ is given by

$$X_m^{i'} = X_m^i + (-1)^n r \,.\, \Delta \tag{1.16}$$

Where r is a random number between 0 and 1, 'n' is a random integer and Δ is a magnitude of change which the coordinate can undergo (supplied). So mutation basically alters the value of a variable by a random magnitude in the positive or negative direction. Mutation in a way is capable of introducing finer features into the solution pool as it operates on individual variables in the trial solutions. This is extremely important as minor adjustments to the magnitude of the variables can only lead the search to a solution which is near to the desired objective. These three operations of Selection, Crossover and Mutation in a particular step completes what is known as a generation. After a given generation the reformed strings are checked for their fitness, arranged in hierarchical manner and again subjected to the same set of operations. This process is continued iteratively until a constant high value of fitness is found or the fitness of the best string is 1. The string with the best fitness is the desired solution and the co-ordinates present in it are the optimized coordinates of the atoms in the cluster. The literature concerning applications of GA in the field of physics and chemistry is gradually increasing. Most of the works are in geometry optimization of cluster, crystals and bio molecules. Typical classic applications of GA are investigations of medium-sized and large clusters using the Born-Mayer[70], vander-waals[71], TIP4P[72] or Gupta potentials[73]. There has been numerous works relating to molecular cluster [74-78], metallic cluster [79-90], Noble gas cluster [91-93] etc. Using an exponential scaling of the Potential Energy Surface (PES) to obtain fitness function for GA, Hartke et al. [94] found global minimum energy structure of Si_4 cluster on an empirical PES. They also used GA based method to search global minimum structures of $Na^+(H_2O)_n$ cluster system with n=4-25[95]. The entire process of Genetic Algorithm is outlined in the flowchart given below.

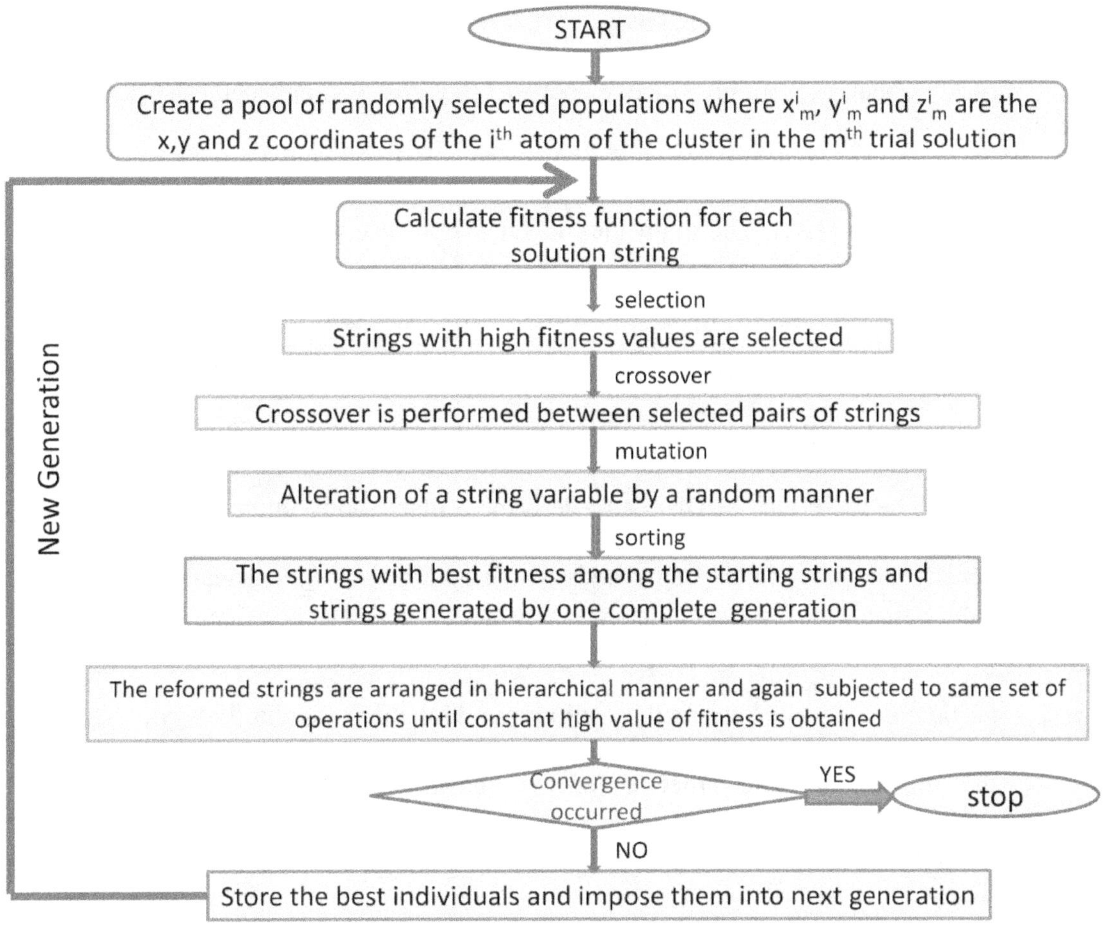

Deaven et al. [97,98] used a novel crossover technique in GA to obtain global minimum structure of LJ cluster of sizes upto 110 atoms. In other approach GA has been used to find global minimum structures of Xe cluster, mixed Xe-Ar cluster [99] and in construction of reaction path on PES for LJ clusters. GA as a structure optimizer has greater contribution to understand the phenomenon of protein folding of biomolecules [100-102]. It has been used in drug designing [103] and pulse shape designing and optimization in experimental spectroscopy [104-108]. In theoretical chemistry GA has been used in solving eigenvalue problem [109], schrodinger equation [110-121] and basis set optimization. Now the application of GA is gradually increasing in different field of study.

1.3 Density Functional Theory

Density functional theory is one of the most popular methods for doing calculations on atoms and molecules, and their interactions [122], and it has an excellent performance to cost ratio [123]. The main idea in DFT is to use the electron charge density n(r) as a basic variable, instead of many electron wave function used in Hartree-Fock theory. DFT has the advantage of being able to solve many problems with high accuracy together with being computationally simple. The scope of this section is to give a short introduction to DFT. The main idea of DFT is to describe a many-body interacting system via its particle density and not via its many-body wave function. Its significance is to reduce the 3N degrees of freedom of the N-body system to only three spatial coordinates through its particle density. Its basis is the well known Hohenberg-Kohn theorem [124], which claims that the ground state molecular energy and all other molecular electronic properties are uniquely determined by the ground-state electron probability density ρ_o(x,y,z), a function of only three variables.

1.3.1 The Hohenberg-Kohn Theorems

The entire field of DFT rests upon the idea that there is a relationship between the total electronic energy and the overall electron density, n(r). The basic idea is that we can use the electron density as a variable instead of many electron wave function as in Hartree-Fock theory, and thus consider all the electrons at once instead of using single particle approach with single-particle orbitals. Hohenberg and Kohn formulated two theorems:

Theorem 1. The ground state energy of a system of interacting electrons is a unique functional of the electronic charge density. The total energy functional is then written as

$$E_v[n] = \int n(r)v(r)dr + F_{HK}[n] \tag{1.17}$$

where $$F_{HK}[n] = T[n] + E_{ee}[n] \tag{1.18}$$

Here $T[n]$ is the kinetic energy and $E_{ee}[n]$ is the energy due to the electron-electron interaction.

Theorem 2. The second H-K theorem states that, $F_{HK}[n]$ the functional that delivers the ground state energy of the system, delivers the lowest energy if and only if the input density is the true ground state density. If we denote $E_v[n_0]$ as the ground state energy and $n_0(r)$ as the ground state electron density, the second Hohenberg-Kohn theorem provides the variational principle. For a trial electron density, such that and, $\int n_{tr}(r)dr = N_e$ then

$$E_v[n_{tr}] = \int v(r)n_{tr}(r)dr + F_{HK}[n_{tr}] \geq E_v[n_0] \quad (1.19)$$

with $$\int v(r)n_0(r)dr + F_{HK}[n_0(r)] = E_v[n_0] \quad (1.20)$$

Assuming differentiability of $E_v[n]$, the variational principle requires that the ground state density satisfies the stationary principle.

$$\delta\{E_v[n] - \mu[\int n(r)dr - N_e]\} = 0 \quad (1.21)$$

Which gives Euler-Lagrange equation $\mu = \frac{\delta E_v[n]}{\delta n(r)} = v(r) + \frac{\delta F_{HK}[n]}{\delta n(r)}$ (1.22)

where μ is the Lagrange multiplier. If the exact form of the functional $F_{HK}[n]$ would be known, then equation (1.21) would be the exact equation for the ground-state electron density. Once we have an explicit form (approximate or accurate) for $F_{HK}[n]$, we can apply this method for any system. After the proposal of the density-functional theory, Kohn and Sham proposed a scheme to treat the variational problem of Eq.(1.21), which is given in the following section.

1.3.2 The Kohn-Sham Equations

We have seen that the ground state energy of a many electron system can be obtained by minimizing the functional

$$E_v[n] = \int v(r)n(r)dr + F_{HK}[n] \quad (1.23)$$

with $$F_{HK}[n] = T[n] + E_{ee}[n] \tag{1.24}$$

In a trade for simplicity, Kohn and Sham [125] proposed an indirect approach to the kinetic energy functional $T[n]$. They proposed to introduce orbitals into the problem in such a way that the kinetic energy can be computed simply to good accuracy. The functional $F_{HK}[n]$ which is unknown until this point, is rewritten as

$$F_{HK} = T_s[n] + J[n] + E_{xc}[n] \tag{1.25}$$

$$J[n] = \frac{1}{2}\int\int \frac{n(r)n(r')}{|r-r'|} dr dr' \tag{1.26}$$

Where $J[n]$ is the classical electrostatic repulsion (Hartree term), $T_s[n]$ is the kinetic energy of a system of non-interacting electrons and $E_{xc}[n]$ is the so-called exchange correlation energy, thus the energy functional $E_v[n]$ becomes

$$E_v[n] = T_s[n] + J[n] + E_{xc}[n] + \int v(r)n(r)dr \tag{1.27}$$

$$= \sum_i^{Nocc} \int \psi_i^*(r)\left(-\frac{1}{2}\nabla^2\right)\psi_i(r)dr + J[n] + E_{xc}[n] + \int v(r)n(r)dr \tag{1.28}$$

where $\psi_i(r)$ are the Kohn—Sham (KS) orbitals and the sum run over all the occupied states N_{occ}. According to Kohn—Sham formalism Kohn—Sham orbitals should be orthonormal.

$$\int \psi_i^*(r)\psi_i(r)dr = \delta_{ij} \tag{1.29}$$

Now let us assume that the functional $\Omega[\{\psi_i\}]$ of N_{occ} orbitals where ε_{ij} are the lagrange multipliers for the constraints of equation (1.29)

$$\Omega[\{\psi_i\}] = E_v[n] - \sum_i^{Nocc}\sum_j^{Nocc} \varepsilon_{ij}\int \psi_i^*(r)\psi_i(r)dr \tag{1.30}$$

For $E_v[n]$ to be a minimum it is necessary that $\delta\Omega[\{\psi_i\}] = 0$

which leads to the equations

$$h_{eff}\psi_i = \left[-\frac{1}{2}\nabla^2 + v_{eff}(r)\right]\psi_i(r) = \sum_j^{N_{occ}} \varepsilon_{ij}\,\psi_i(r) \qquad (1.31)$$

with $v_{eff}(r)$ determined through

$$v_{eff}(r) = v(r) + \frac{\delta J[n]}{\delta n(r)} + \frac{\delta E_{xc}(n)}{\delta n(r)}$$

$$= v(r) + \int \frac{n(r')}{|r-r'|} dr + v_{xc}(r)$$

$$= v(r) + v_H(r) + v_{xc}(r) \qquad (1.32)$$

where $v_{xc}(r)$ is the so-called exchange-correlation potential. Therefore, for a given $v_{eff}(r)$ one obtains n(r) that satisfies the equation (1.33) by solving the one-electron equations

$$\left[-\frac{1}{2}\nabla^2 + v_{eff}(r)\right]\psi_i(r) = \varepsilon_i\psi_i(r) \qquad (1.33)$$

and setting

$$n(r) = \sum_i^{N_{occ}} |\psi_i(r)|^2 \qquad (1.34)$$

The equations (1.32), (1.33), (1.34) are the so-called Kohn-Sham equations and need to be solved self-consistently. It follows then, that the total energy is calculated as

$$E_e = \sum_j^{N_{occ}} \varepsilon_i - \frac{1}{2}\int \frac{n(r)n(r')}{|r-r'|} dr dr' + E_{xc}[n] - \int v_{xc}(r)n(r)dr \qquad (1.35)$$

1.3.3 Approximations for exchange correlation functional

To find the ground state energy, we have to solve the Kohn-Sham equations, i.e. to minimize an energy functional and self-consistently find the solutions to single-particle equations. To be able to do this, we have to specify the exchange-correlation functional. The real form of this functional is unfortunately not known. This means that approximations must be made.

1.3.3.1 Local Density Approximation (LDA)

The most fundamental and simplest approximation is the local-density approximation (LDA), in which the energy depends only on the density at the point where the functional is evaluated. For this case the electron density is constant in all space. The easiest approximation to the exchange correlation functional uses this case to construct a Kohn-Sham exchange correlation functional. If we assume that the exchange correlation functional at each point is the exchange correlation functional for the uniform electron gas given the electron density at that point, we get

$$V_{XC}^{LDA}(r) = V_{XC}^{gas}[n(r)] \tag{1.36}$$

This is the Local Density Approximation (LDA), since it only uses the local density to construct an exchange correlation functional. The LDA fits well with bulk materials where the electron density is slowly varying. The electron density for atoms and molecules does not have this property, which means that the LDA is not the best choice for predicting properties in these cases.

1.3.3.2 Generalized Gradient Approximation (GGA)

A second class of exchange correlation functionals also uses the local electron density as in LDA, but this class also includes the local gradient in the electron gas. Generalized gradient approximation (GGA) incorporates dependence not only on the electron density but also on its gradient, thus being able to better describe the inhomogeneous nature of molecular densities. This is done because real electron densities are not uniform, so by including information about the spatial variation the functional can be improved. This approximation is called the Generalized Gradient Approximation (GGA):

$$V_{XC}^{GGA}(r) = V_{XC}[n(r), \nabla n(r)] \quad (1.37)$$

BLYP is an example of a GGA exchange correlation functional, and it combines the Becke gradient-exchange correction with the Lee-Yang-Parr correlation functional [126]. Typical molecular systems are generally very different from a homogenous electron gas. In fact any real system is spatially inhomogeneous. Generalized gradient approximation (GGA) methods take into account this effect, by making the exchange and correlation energies dependent not only on the density but also on the gradient of the density $\Delta n(r)$[127-132]. Thus, GGA extends the practical implementation of DFT to inhomogeneous systems and credited as nonlocal method. GGA methods represent a significant improvement over the local methods and it tends to give better total energies, atomization energies, structural energy differences, and reaction barriers for molecules [133-136]. Although the GGA includes more properties than the LDA, it does not always give better results. But it generally does so for small molecules and atoms.

1.3.3.3 Hybrid functional

The Kohn-Sham exchange-correlation functional includes the exchange and the correlation effects, which both are unknown. Hartree-Fock theory however, has an exact expression for the exchange energy. Thus, many attempts in coupling HF theory and DFT have been done, and in 1992 Becke [137] was the first that succeeded. Since then, many mixed functionals have been developed. All functional that mix a part of the nonlocal HF exact exchange with a DFT exchange-correlation functional are called hybrid functionals. The most popular hybrid functional is the B3LYP functional [138]. The B3LYP functional can be written as

$$V_{XC}^{B3LYP} = V_{XC}^{LDA} + \alpha_1(E_X^{HF} - V_X^{LDA}) + \alpha_2(V_X^{GGA} - V_X^{LDA}) + \alpha_3(V_C^{GGA} - V_C^{LDA}) \quad (1.38)$$

where V_X^{GGA} is the Becke 88 exchange functional and V_C^{GGA} is the Lee-Yang-Parr correlation functional, the same functionals that together make up the BLYP functional mentioned above. E_X^{HF} is the HF exact exchange, and α_1, α_2 and α_3 are numerical parameters. As we can see, the B3LYP functional combines exchange and correlation functionals from both LDA and GGA with the exact exchange from HF. The B3LYP functional, together with some other hybrid functionals, has been highly successful in predicting properties of small molecules. A big advantage with DFT, is that it has low computational costs. However, when HF exact exchange is added, the computational cost is greatly increased.

1.4 Atomic and Molecular clusters

Cluster Science has become an important area of research for both experimentalist and theoreticians in the last two decades. Both atomic and molecular clusters with diverse bonding patterns among the constituents are studied and characterized with increasing effort. Scientists look into these chemical systems from diverse angles. In atomic clusters, especially noble gas clusters, huge amount of literature exists with studies on structural shapes in size specific clusters, deciphering of magic numbers, studying Coulomb explosions in charged clusters and determining critical sizes for stability and identification of dissociation channels. There has also been considerable work on mapping out reaction paths and locating Transition state structures in both neutral and doped rare gas clusters. Theoreticians have done their bit in these fields also. Simulations using Classical Molecular Dynamics [139], Ab-initio Molecular Dynamics[140], Monte-Carlo Simulated Annealing [141] and Genetic Algorithm[142] have enriched the existing literature in this field. Metallic Clusters have also got their share of deserved attention. Both neutral and metal-metal doped clusters have been studied with vigour [143, 144]. These systems are also important constituents as nano-catalysts. Metal clusters in their finely divided state serve as potent catalysts in many organic reactions and there is every chemical reason to believe that cluster of nano-size range, because of their enhanced surface area, can serve as even better catalysts. Study of atomic and molecular clusters is a challenging field to be engaged in as it throws up new and interesting features for experimentalists to find out and for theoreticians to provide logical explanations consistent with experimental findings. The area of hydrated halide ion clusters is an important system for study from the point of view of bonding, structure and spectroscopy. Experimentally the creation of size specific hydrated halide ion clusters have become a reality, and once such clusters are produced, one can go on to measure its spectroscopic properties. The work of Johnson et al. deserves a special mention in this context, especially the detail study of a solute halide ion interacting and forming cluster with a single solute molecule. A simple system like $(H_2O)_nX^-$ (X^- = halide ion) can show up spectroscopic details which can be a challenge for any theoretical model or method to explain in the right perspective[145,146]. The structural properties of these systems can become extremely complicated, with the appearance of many local minima other than the usually sought after global one. The structures especially when the number of solvated water molecules are increased, show two types of hydrogen bonding, the O—H----X^- and O—H----O type. These hydrogen bonds also leave their mark on the IR spectroscopy of these systems. CO_2 gas plays very important role in respiration and photosynthesis which are responsible for the existence of life on earth. CO2 finds its application in greenhouses as the main carbon source. Due to widespread applications, CO_2 clusters have been a subject of major experimental investigations[147–157]. Among the molecular clusters CO_2 rank second only to H_2O due to the importance of atmospheric CO_2 for the earth as well as the advantages of super-critical CO_2 as an industrial solvent[158,159]. There has been a persistence effort from both experimentalists and theoreticians to explain and understand the structural and spectroscopic properties of $(CO_2)_n$ clusters. Liu et al.[160] applied Monte-Carlo simulation technique in studying CO_2 clusters. Geometry Optimization of carbon dioxide clusters i.e; $[(CO_2)_n]$ for $4 \leq n \leq 40$ was performed by Takeuchi[161] using Lennard Jones (LJ) potential. Yeole et al.[162] have used Molecular Tailoring approach(MTA) to study structure, energetics and vibrational spectra of CO_2 clusters. Liu and Jordan[163] employed parallel-tempering Monte Carlo method to investigate finite temperature properties of $(CO_2)_n$ clusters. Very recently Temke et al.[164] studied thermodynamic properties of carbon dioxide Clusters by M06-2X and Dispersion-Corrected B2PLYP-D theory. Nanoscale clusters[165-167] are an important topic in current scientific research because nanotubes, rods and cages are interesting structural motifs and are candidates for electrical and mechanical parts in nanodevices. Hence studies of cluster properties as a function of size has attracted researchers in recent years. The study of the properties of metal-oxide nanostructures has emerged as a research field of great interest in recent times. Magnesium oxide (MgO) is a simple metal oxide which is widely used in technological applications. Bulk MgO is an inert material with a high melting point and is also a wide bandgap(7.8eV)insulator. Saunders[168,169] studied the stability of (MgO)3 clusters by performing experiments in quadrupole mass spectrometry. Size selective infrared spectra of (MgO)n clusters with n=3-16 have been studied in gas phase with a tunable IR-UV two color ionization scheme by Haertelt et al[170]. Moukouri and Nogurea[171] have performed semi-empirical tight binding calculations on (MgO)n clusters with n=1-6. Quantum Chemical methods[172-174] are used by theoreticians to study size dependent structural and electronic properties of $(MgO)_n$ clusters. Zhang et al.[175] employed molecular dynamics simulation to study $(MgO)_n$ clusters with n=18, 21 and 24. Dong et al.[176] explored the stability of $(MgO)_{3n}$ (n=2-10) clusters using Genetic Algorithm in association with Density Functional Theory(DFT). Transition metal oxide clusters[177-189] have attracted the attention of researchers due to their widespread application in catalysis. Structure and electronic properties of TiO_2 nanoparticles is an active field of research owing to its application in photocatalysis and photovolatics. Clusters of titanium oxide are very important for their widespread application [190-199] in diverse areas of chemistry, physics and material science. Titania(TiO_2) is the fundamental material in the field of gas-sensors, photovolatics, photocatalysis, energy storage. The semiconductor property of titania (TiO_2) find its important activity in solarcells[200], photocatalysis[201,202], bacterial inactivation[203], water disinfection[204], sunscreens[205]. Hamad et al.[206] reported various structures of TiO_2 clusters using pair potential based simmulated annealing Monte Carlo basin hopping (SA-MCBH) and genetic algorithm(GA) methods. Coupled-cluster calculations involving single and double excitations was used by Li and Dixon[207]. Density functional calculations on (TiO_2)n clusters were performed by Shevlin et al.[208]. Qu and kroes[209, 210] studied TiO2 clusters using B3LYP/LANL2DZ method and obtained lower energy structures. Structure and reactiviy of small size $(TiO_2)_{1-10}$ has been investigated by Calatayud et al.[211, 212]. Chiodo et al.[213] investigated structural, electronic and optical properties of (TiO2)n clusters using ab initio DFT calculation. Mowbray et al.[214] investigated Stability and Electronic Properties of TiO_2 Nanostructures with and without B and N doping. Very recently Rana et al.[215] performed abinitio study to understand stability and magnetism of $(TiO_2)_n$ clusters.

1.5 Plan of the thesis

The main objective of this thesis is to discuss stochastic optimization technique combined with quantum chemical method (DFT) on solving chemical problems (related to cluster). One of the prime objective is to find for a given cluster size the arrangement of atoms corresponding to the lowest potential energy i.e the global minimum (GM) on the potential energy hypersurface. As the number of basins on the energy hypersurface will increase dramatically with the number of atoms, so the probability that any LM found is the GM will fall dramatically. Thus global optimisation techniques are typically used to generate several plausible structures that can be further refined in the second stage. In fact , with the optimisation techniques discussed above, the task is to find any point within each of the respective basins that contain either the global minimum or the low energy metastable local minima. Thus, interatomic potential methods can be used when comparing structures during the first stage, the global search, and quantum chemical method (DFT) in stage two. In the second stage quantum chemical method can be employed to test the reliability of the first stage calculations by probing cluster structures, which are likely to be sensitive to the parameterisation of the interatomic potential used. We can thus obtain and predict not only the atomic structure, but also gain insight into the electronic structure of these nanoparticles and extend these studies further to the physical and chemical properties of the materials, including optical and magnetic response which would allow a direct comparison with the available experimental spectroscopic data. In this thesis our objective was to see if the strategy used by us of combining an initial algorithmic search, which is stochastic, followed by adequate quantum chemical DFT study using the output obtained from the natural stochastic evolution is a viable strategy or not for evaluating global minimum as well as low energy structures of different types of cluster systems. To examine in depth the efficiency of the suggested strategy, we have gone ahead to study cluster systems formed by different types of interacting forces like the strong coulombic interaction as well as sufficient degree of covalent interactions in certain systems. To further check, if our strategy is equally applicable in situations where the interacting chemical force is weak in nature, we have also studied cluster systems formed by purely hydrogen bonds as well as dispersion forces. This diverse range of chemical application handled by our strategy [the only common element in all these chemical systems is that they are all clusters] suggest that the proposed method has an inherent power to handle diverse chemical situations and application. In the chapters that follow the following cluster systems have been handled. In chapter 2 we examine $(H_2O)_nF^-$ cluster system, where the interaction is hydrogen bonded. The encapsulated fluoride ion being relatively small in size the structures generated are mostly of an encapsulated type. In chapter 3 we again examine H-bonded cluster systems, specifically $(H_2O)_nX^-$ [X=Cl, Br and I]. These systems broadly prefer surface capped structures owing to the larger size of the ion as opposed to $(H_2O)_nF^-$. In chapter-4 we study the $(CO_2)_n$ cluster system where the interacting force is still weak but it is Vander Waals type as opposed to hydrogen bonded system. Chapter-5 deals with MgO system and here the interaction force is pretty strong as it is predominantly coulombic. Chapter-6 which is the last cluster system to be discussed in the thesis, deals with the $(TiO_2)_n$ system where the interaction forces have a large degree of covalent character.

1.6 References

[1] R. L. Johnston, Atomic and Molecular Clusters, Taylor and Francis, London, **2002**

[2] J. Jortner, Z. Phys. D, **1992**, 24, 247

[3] R. L. Johnston, Philos. Trans. R. Soc. London, Ser. A, **1998**, 356, 211

[4] T. P. Martin, Phys. Reports **1996**, 273, 199

[5] L. D. Marks, Reports Prog. Phys. **1994**, 57, 603

[6] M. Moseler, H. H€akkinen, R. N. Barnett, U. Landman, Phys. Rev. Lett. **2001**, 86, 2545.

[7] D. J. Wales, H. A. Scheraga, Science **1999**, 285, 1368

[8] Theory of Atomic and Molecular Clusters, ed. J. Jellinek, Springer, Berlin, **1999**

[9] J. P. K. Doye, Phys. Rev. B **2003**, 68, 195418

[10] M. D. Wolf, U. Landman, J. Phys. Chem. A **1998**, 102, 6129

[11] B. Huber, M. Moseler, O. Kostko, B. v. Issendorff, Phys. Rev. B **2009**, 80, 235425

[12] A. A. Sokol, C. R. A. Catlow, M. Miskufova, S. A. Shevlin, A. A.nAl-Sunaidi, A.Walsh, S. M. Woodley, Phys. Chem. Chem. Phys. **2010**, 12, 8438

[13] S. M. Woodley, C. R. A Catlow, Nature Mater. **2008**, 7, 937

[14] S. M. Woodley, S. Hamad, C. R. A. Catlow, Phys. Chem. Chem. Phys. **2010**, 12,8454

[15] C. R. A. Catlow, S. T. Bromley, S. Hamad, M. Mora-Fonz, A. A. Sokol, S.M.Woodley, Phys. Chem. Chem. Phys. **2010**, 12, 786

[16] W. Q. Huang, X. J. Lai, R. C. Xu, Chem. Phys. Lett. **2011**, 507, 199

[17] X. J. Lai, W. Q. Huang, R. C. Xu, J. Phys. Chem. A **2011**, 115, 5021

[18] S. M. Woodley and R. Catlow, Nat. Mater., **2008**, 7, 937–946

[19] S. M. Woodley, in Applications of Evolutionary Computation in Chemistry,Springer-Verlag Berlin, Berlin, **2004**, vol. 110, pp. 95–132

[20] B. Hartke, J. Comput. Chem., **1999**, 20, 1752

[21] Z. Y. Li, N. P. Young, M. Di Vece, S. Palomba, R. E. Palmer, A. L. Bleloch, B. C. Curley, R. L. Johnston, J. Jiang, J. Yuan, Nature **2008**, 451, 46.

[22] D. Schooss, P. Weis, O. Hampe, M. M. Kappes, Phil. Trans. R. Soc. A **2010**, 368, 1211.

[23] T. Udayabhaskararao, Y. Sun, N. Goswami, S. K. Pal, K.Balasubramanian,T.Pradeep, Angew. Chem. Int. Ed. **2012**, 51, 2155.

[24] J. P. K. Doye, D. J. Wales, J. Phys. Chem. A **1997**, 101, 5111.

[25] R. H. Leary, J. P. K. Doye, Phys. Rev. E **1999**, 60, R6320.

[26] R. H. Leary, J. Global Optim. **2000**, 18, 367.

[27] L. X. Zhan, J. Z. Y. Chen, W. K. Liu, S. K. Lai, J. Chem. Phys. **2005**, 122, 244707.

[28] H. G. Kim, S. K. Choi, H. M. Lee, J. Chem. Phys. **2008**, 128, 144702.

[29] A. Uhlherr, Comput. Phys. Commun. **2003**, 155, 31.

[30] S. Hayryan, C. K. Hu, S. Y. Hu, R. J. Shang, J. Comput. Chem. **2001**, 22, 1287.

[31] D. J. Adams, J. Comput. Phys. **1988**, 75, 138.

[32] X. G. Shao, L. J. Cheng, W. S. Cai, J. Comput. Chem. **2004**, 25, 1693.

[33] X. L. Yang, W. S. Cai, X. G. Shao, J. Comput. Chem. **2007**, 28, 1427.

[34] X. G. Shao, X. L. Yang, W. S. Cai, J. Comput. Chem. **2008**, 29, 1772.

[35] X. Wu, W. S. Cai, X. G. Shao, Chem. Phys. **2009**, 363, 72.

[36] L. Davis, Handbook of Genetic Algorithms; Van Nostrand Reinhold: NewYork,**1991**.

[37] J. R. Morris, D. M. Deaven, K. M. Ho, Phys. Rev. B **1996**, 53, 1740.

[38] B. Hartke, J. Phys. Chem. **1993**, 97, 9973.

[39] Y. Xiao, D. E. Williams, Chem. Phys. Lett. **1993**, 215, 17.

[40] Y. Zeiri, Phys. Rev. E **1995**, 51, 2769.

[41] Y. Zeiri, J. Phys. Chem. A **1998**, 102, 2785.

[42] D. M. Deaven, K. M. Ho, Phys. Rev. Lett. **1995**, 75, 288.

[43] D. M. Daven, N. Tit, J. R. Morris, K. M. Ho, Chem. Phys. Lett. **1996**, 256,195.

[44] W. J. Pullan, J. Chem. Inf. Comput. Sci. **1997**, 37, 1189.

[45] C. Roberts, R. L. Johnston, N. T. Wilson, Theor. Chim. Acta **2000**, 104,123.

[46] R. L. Johnston, Dalton Trans. **2003**, 22, 4193.

[47] O. O~na, V. E. Bazterra, M. C. Caputo, J. C. Facelli, P. Fuentealba, M. B. Ferraro, Phys. Rev. A **2006**, 73, 053203.

[48] D. X. Tian, H. L. Zhang, J. J. Zhao, Solid State Commun. **2007**, 144, 174.

[49] J. Rogan, G. Garcı´a, M. Ramı´rez, V. Mu~noz, J. A. Valdivia, X. Andrade, R. Ramı´rez, M. Kiwi, Nanotechnology **2008**, 19,205701.

[50] K. Oleksy, F. Karlicky´, R. Kalus, J. Chem. Phys. **2010**, 133, 164314.

[51] J. L. Llanio-Trujillo, J. M. C. Marques, F. B. Pereira, J. Phys. Chem. A **2011**, 115, 2130.

[52] C. Gollub, R. de Vivie-Riedle, Phys. Rev. A **2009**, 79, 21401.

[53] Y. C. Wang, J. Lv, L. Zhu, Y. M. Ma, Phys. Rev. B **2010**, 82, 094116.

[54] W. S. Cai, H. Y. Jiang, X. G. Shao, J. Chem. Inf. Comput. Sci. **2002**, 42, 1099.

[55] S. Kirkpatrick, C. D. Gelatt, M. P. Vecchi, Science **1983**, 220, 671.

[56] S. Talukder, P. Chaudhury, R. Metzler, S. K. Banik J. Chem. Phys **2011**, 135,165103

[57] S. Guha, P. Chaudhury THEOCHEM **2010**, 945, 12

[58] S. K. Biring, P. Chaudhury Chem. Phys. **2010**, 377, 46

[59] S. Guha, P. Chaudhury Struct. Chem **2011**, 22, 1007

[60] S. K. Biring, P. Chaudhury Chem. Phys. **2012**, 400, 198

[61] S. Guha, N. Mukherjee, P. Chaudhury Indian. J. Phys **2012**, 86, 245

[62] M. Mihalkovic, C. L. Henley Phys. Rev. B **2013**, 88, 064201-1

[63] P. N. Samanta, K. K. Das Comput. Theo. Chem **2012**, 1000, 42

[64] P. Dutta, S. P. Bhattacharya Phys. Lett. A **1990**, 148, 331

[65] P. Dutta, D. Mazumdar, S. P. Bhattacharya Chem. Phys. Lett. **1991**, 181, 67

[66] P. Dutta, S. P. Bhattacharya Chem. Phys. Lett **1991**, 184, 330

[67] P. Chaudhury, P. Dutta, P. Bandopadhyay, P. Sarkar, S. P. Bhattacharya Chem.Phys. Lett. **1996**, 250, 238

[68] P. Chaudhury, S. P. Bhattacharya THEOCHEM **1998**, 429, 175

[69] K. F. C. Yiu, K. Y. Tam, S. C. Tsang J. Comput. Chem **1997**, 18, 290

[70] C. Roberts, R. L. Johnston, Phys. Chem. Chem. Phys **2001**, 3, 5024

[71] B. Hartke, J. Comput. Chem. **1999**, 20, 1752

[72] B. Hartke Z. Phys. Chem **2000**, 214, 1251

[73] S. Darby, T. V. Mortimer-Jones, R. L. Johnston, C. Roberts, J. Chem. Phys. **2002**, 116, 1536.

[74] P. Chaudhury, S. P. Bhattacharya Chem. Phys. **2001**, 270, 277

[75] S. G. Neogi, P. Chaudhury Ind. J. Phys. **2014**, 88, 781

[76] S. G. Neogi, P. Chaudhury Struc. Chem **2014**, 25, 909

[77] S. G. Neogi, P. Chaudhury J. Comput. Chem **2013**, 34, 471

[78] S. G. Neogi, P. Chaudhury J. Comput. Chem **2012**, 33, 629

[79] P. C. Jennings, R. L. Johnston Comput. Theo. Chem **2013**, 1021, 91

[80] X. Huang, L. Sai, X. Jiang, J. Zhao Euro. Phys. J. D **2013**, 67, 43

[81] C. J. Heard, R. L. Johnston Euro. Phys. J. D **2013**, 67, 34

[82] M. Chen, D. A. Dixon J. Chem. Theo. Comput. **2013**, 9, 3189

[83] A. Erlebach, C. Hunn, R. Jana, M. Sierka Phys. Chem. Chem. Phys **2014** DOI 10.1039/c4cp02099e

[84] M. X. Silva, B. R. L. Galvo, J. C. Belchior Phys. Chem.Chem. Phys **2014**, 16, 8895

[85] S. G. Neogi, P. Chaudhury J. Comput. Chem **2014**, 35, 51

[86] R. L. Johnston Dalton Trans **2003**, 4193

[87] S. Heiles, A. J. Logsdial, R. Schafer, R. L. Johnston Nanoscale **2012**, 4, 1109

[88] J. B. A. Davis, S. L. Horswell, R. L. Johnston J. Phys. Chem. A **2014**, 118, 208

[89] S. Heiles, R. L. Johnston Int. J. Quant. Chem **2013**, 113, 2091

[90] C. J. Heard, R. L. Johnston Phys. Chem. Chem. Phys **2014**,DOI:10.1039/c3cp55507k

[91] P. Chaudhury, S. P. Bhattacharya Ind. J. Phys. B **1999**, 73, 191

[92] J. Zhao, R. H. Xie J. Comput. Theo. NanoSc **2004**, 1, 117

[93] A. Stasikowski, M. Moneta, T. M. Gwizdalla Surf. and Intrf. Analysis **2006**, 38,469

[94] B. Hartke, J. Phys. Chem **1993**, 97, 9973

[95] B. Hartke, A. Charvat, M. Reich, B. Abel, J. Chem. Phys **2002**, 116, 3588

[96] P. Chaudhury, R. Saha, S. P. Bhattacharya Chem. Phys. Lett **2001**, 270, 277

[97] D. M. Deaven, N. Tit, J. R. Morris, K. M. Ho Phys. Rev. Lett **1995**, 75, 288

[98] D. M. Deaven, N. Tit, J. R. Morris, K. M. Ho Chem. Phys. Lett **1996**, 256, 195

[99] P. Chaudhury, S. P. Bhattacharya, Ind. J. Phys. B, **1999**, 73, 191

[100] A. Piccolboni, G. Mauri, Artificial Evolution **1998**, 136, 3123

[101] T. Z. Jiang, Q. H. Cui, G. H. Shi, S. D. Ma J. Chem. Phys **2003**, 119, 4592

[102] D. De. Sancho, L. Prieto, A. M. Rubio, A. Rey J. Comput. Chem **2005**, 26, 131

[103] E. W. Lameijer, T. Baeck, J. N. Kok, A. P. Ijzerman Natural Computing **2005**, 4,177

[104] A. Efimov, M. D. Moores, B. Mei, J. L. Krause, C. W. Siders, D. H. Reitze App.Phys. B – Lasers and Optics, **2000**, 70, S133

[105] A. Dargys, Ultrafast Phenomenon in Semiconductors **2002**, 384, 3259

[106] R. Mizoguchi, K. Onda, S. S. Kano, A. Wada, Review of Scientific instruments **2003**, 74, 2670

[107] M. Tsubouchi, T. Momose, Phys. Rev. A, **2008**, 77, 052326

[108] I. F. Barna, P. Dombi Central European Journal of Physics **2008**, 6, 205

[109] P. Chaudhury, S. P. Bhattacharya Chem. Phys. Lett **1998**, 296, 51

[110] H. Nakanishi, M. Sugawara Chem. Phys. Lett **2000**, 327, 429

[111] M. Sahin, M. Tomak, Int. J. Mod. Phys. C **2002**, 26, 3883

[112] H. Safak, M. Sahin, B. Gulveren, M. Tomak, Int. J. Mod. Phys. C **2003**, 14, 775

[113] M. Sahin, U. Atav, M. Tomak, Int. J. Mod. Phys. C **2005**, 16, 1379

[114] M. Sahin, M. Tomak, Physica E. **2005**, 28, 247

[115] I. Grigenko, M. E. Gracia, Physica A. **2000**, 284, 131

[116] I. Grigenko, M. E. Gracia, Physica A. **2000**, 291, 439

[117] I. Grigenko, M. E. Gracia, Physica A. **2000**, 313, 463

[118] I. Grigenko, O. Speer, M. E. Garcia, Phys. Rev. B **2002**, 65, 235309

[119] R. Saha, P. Chaudhury, S. P. Bhattacharya, Phys. Lett. A, **2001**, 291, 397

[120] R. Saha, S. P. Bhattacharya, J. Theo. Comput. Chem, **2004**, 3, 325

[121] R. Saha, S. P. Bhattacharya, Int. J. Mod. Phys. C **2007**, 18, 163

[122] N. Argaman and G. Makov, "Density functional theory – An introduction," American Journal of Physics, **2000**, 68, 69-79

[123] Y. Zhao and D. Truhlar, "Density Functionals with Broad Applicability in Chemistry,'' Accounts of Chemical Research **2008**, 41, 157

[124] P. Hohenberg and W. Kohn, Phys. Rev. B **1964**, 136, 864

[125] W. Kohn and L. J. Sham, Phys. Rev. A **1965**, 140, A1133

[126] A. Leach, Molecular modelling: principles and applications Pearson Education Limited, 2nd ed. (**2001**)

[127] A. D. Becke, J. Chem. Phys **1986**, 84, 4524

[128] J. P. Perdew, Phys. Rev. B **1986**, 33, 8832

[129] J. P. Perdew and Y. Wang, Phys. Rev. B **1992**, 45, 13244

[130] J. P. Perdew, K. Burke and M. Ernzerhof, Phys. Rev. Lett. **1996**, 77, 3865

[131] A. D. Becke, J. Chem. Phys. **1988**, 88, 1053

132] C. Lee, W. Yang and R. G. Parr, Phys. Rev. B **1988**, 37, 785

[133] A. D. Becke, Phys. Rev. A **1988**, 38, 3098

[134] A. D. Becke, J. Chem. Phys. **1992**, 96, 2155

[135] D. C. Langreth and M. Mehl, Phys. Rev. B **1983**, 28, 1809

[136] J. P. Perdew, J. A. Chevary, S. H. Vosko, K. A. Jackson, M. R. Pederson, D. J.Singh and C. Fiolhais, Phys. Rev. B **1992**, 46, 6671

[137] A. Becke, J. Chem. Phys **1993**, 98, 1372

[138] D. S. Sholl and J. A. Steckel, Density Functional Theory. A practical introduction. Wiley (**2009**)

[139] I. Last, J. Jortner, Phys. Rev. A **2007**, 75, 042507-1

[140] R. Car, M. Parrinello, Phys. Rev. Lett. **1985**, 55, 2471

[141] S. Kirkpatrick, C. D. Gelatt, M. P. Vecchi, Science **1983**, 220, 671

[142] D. E. Goldberg, Genetic Algorithm in Search, Optimization and Machine learning; Addison Wesley: Reading MA, **1989**

[143] J. P. K. Doye, Comput. Mat. Sci. **2006**, 35, 227

[144] L. O. Paz-Borbion, A. Gupta, R. L. Johnston , J. Mat. Chem **2008**, 18, 4154

[145] M. A. Johnson, M. L. Alexander, W. C. Lineberger, Chem. Phys. Lett **1984**, 112, 285

[146] J. R. Roscioli, E. G. Dicken, M. A. Johnson, J. Phys. Chem. A **2006**, 110, 4943

[147] A. S. Gopalan, C. M. Wai, H. K. Jacobs ACS Symposium series **2003**, 860

[148] W. Stafford Nature **2007**, 448, 526

[149] M. A. Walsh, T. H. England, T. R. Dyke, B. J. Howard Chem. Phys. Lett **1987**, 142, 265

[150] K. W. Jucks, Z. S. Huang, D. Dayton, R. E. Miller, W. J. Lafferty J. Chem. Phys **1987**, 86, 4341

[151] G. T. Fraser, A. S. Pine, W. J. Lafferty, R. E. Miller J. Chem. Phys **1987**, 87, 1502

[152] M. J. Wedia, D. J. Nesbit J. Chem. Phys **1996**, 105, 10210

[153] G. Torchet, M-F de Feraudy, A. Boutin, A. H. Fuchs J. Chem. Phys **1996**, 105, 3671

[154] Y. Negishi, T. Nagata, T. Tsukuda Chem. Phys. Lett **2002**, 364, 127

[155] O. Ingolfsson, A. M. Wodtke Chem. Phys. Lett **2002**, 360, 415

[156] A. Bonnamy, R. Georges, E. Hugo, R. Sugnorell Phys. Chem. Chem. Phys **2005**, 7, 963

[157] A. Bonnamy, R. Georges, A. Benidar, J. Boissoles, A. Canosa, B. R. Rowe J.Chem. Phys **2003**, 118, 3612

[158] J. M. Desimone Science **2002**, 297, 799

[159] S. Balasubramanian, A. Kohslmeyer, M. L. Klein J. Chem. Phys **2009**, 131, 144506

[160] H. Liu, K. D. Jordan J. Phys. Chem. A **2003**, 107, 5703

[161] H. Takeuchi J. Phys. Chem. A **2008**, 112, 7492

[162] S. D. Yeole, N. Sahu, S. R. Gadre Phys. Chem. Chem. Phys **2012**, 14, 7718

[163] H. Liu, K. D. Jordan J. Phys. Chem. A **2003**, 107, 5703

[164] K. H. Lemke and T. H. Seward, Chem. Phys. Lett **2013**, 573, 19

[165] S. K. Sahoo, S. Pal, P. Sarkar, C. Majumder, Chem. Phys. Lett, **2011**, 516, 6871

[166] S. Saha, S. Sarkar, S. Pal and P. Sarkar, Rsc. Advan. **2013**, 3, 532

[167] S. Pal, R. Sharma, B. Goswami, P. Sarkar, Chem. Phys. Lett. **2009**, 467, 365

[168] W. A. Saunders, Z. Phys. D **1989**, 12, 601

[169] W. A. Saunders, Phys. Rev. B **1998**, 37, 6583

[170] M. Haertelt, A. Fielicke, G. Meijer, K. Kwapien, M. Sierka, J. Sauer, Phys. Chem. Chem. Phys. **2012**, 14, 2849

[171] S. Moukouri, C. Nogurea, Z. Phys. D. At. Mol. Clus. **1992**, 24, 71

[172] Y. Zhang, H. S. Chen, B. X. Liu, C. R. Zhang, X. F. Li, Y. C. Wang, J. Chem. Phys. **2010**, 132, 194304

[173] L. Chen, C. Xu, X. F. Zhang, Theochem. **2008**, 863, 55

[174] L. Chen, C. Xu, X. F. Zhang, C. A. Cheng, T. Zhou, Int. J. Quant. Chem. **2009**, 109 349

[175] Y. Zhang, H. S. Chen, B. X. Liu, C. R. Zhang, X. F. Li, Y. C. Wang, J. Chem. Phys. **2010**, 132, 194304

[176] R. Dong, X. Chen, X. Wang, W. Lu, J. Chem. Phys. **2008**, 129, 044705

[177] S. Li, H. J. Zhai, L. S. Wang, D. A. Dixon J. Phys. Chem. A **2012**, 116, 5256

[178] S. Li, J. M. Hennigan, D. A. Dixon J. Phys. Chem. A **2009**,113, 7861

[179] S. Li, D. A. Dixon J. Phys. Chem. A **2006**, 110, 6231

[180] S. Li, D. A. Dixon J. Phys. Chem. A **2007**, 111, 11093

[181] S. Li, D. A. Dixon J. Phys. Chem. A **2007**, 112, 11908

[182] S. Li, D. A. Dixon J. Phys. Chem. A **2008**, 112, 6646

[183] H. J. Zhai, B. Kiran, L. F. Cui, X. Li, J. Li, L. S. Wang J. Am. Chem. Soc **2004**, 126, 16134

[184] H. J. Zhai, S. Li, D. A. Dixon, L. S. Wang J. Am. Chem. Soc **2008**, 130, 5167

[185] S. Li, H. J. Zhai, L. S. Wang, D. A. Dixon J. Phys. Chem. A **2009**, 113, 11273

[186] S. Li, D. A. Dixon J. Phys. Chem. C **2011**, 115, 19190

[187] S. Li, L. C. Guenther, M. S. Kelley, D. A. Dixon J. Phys. Chem. C **2011**, 115, 8072

[188] H. J. Zhai, B. Wang, X. Huang, L. S. Wang J. Phys. Chem. A **2009**, 113, 3866

[189] H. J. Zhai, L. S. Wang Chem. Phys. Lett **2010**, 500, 185

[190] M. Gratzel, Nature **2001**, 414, 338

[191] A. Fujishima, K. Honda, Nature **1972**, 238, 37

[192] A. L. Linsebigler, G. Lu, J. T. Jr. Yates, Chem. Rev. **1995**, 95, 735

[193] U. Dicbold, Surf. Sci. Rep. **2003**, 48, 53

[194] A. G. Agrios, P. Pichat, J. Appl. Electrochem. **2005**, 35, 655

[195] M. R. Hoffmann, S. T. Martin, W. Choi, D. Bahnemann, Chem. Rev. **1995**, 95, 69

[196] A. M. Volodin, Catal. Today. **2000**, 58, 103

[197] T. L. Thompson, J. T. Jr. YAtes, Chem. Rev. **2006**, 106, 4428

[198] A. Hagfeldt, M. Gratel, Chem. Rev. **1995**, 95, 49

[199] Y. Wang, N. Herron, J. Phys. Chem. **1991**, 95, 525

[200] V. M. Aroutiounian, V. M. Arakelyan, G. E. Shahnazaryan, Sol. Energy. **2005**, 78, 581

[201] H. Iiire, K. Sunanda, K. Hashimoto, Electrochemistry. **2004**, 72, 807

[202] D. Bahnemann, Sol. Energy., **2004**, 77, 445

[203] A. Vohra, D. Y. Goswami, D. A. Deshpande, S. S. Block J. Ind. Microbiol. Biotechnol. **2005**, 32, 364

[204] P. Fernandez, J. Blanco, C. Sichel, S. Malato, Catal. Today. **2005**, 101, 345

[205] V. A. Schwarz, S. D. Klein, R. Hornung, R. Knochenmuss, P. Wyss, D. Fink, U. Haller, H. Watt, Lasers. Surg. Med. **2001**, 29, 252

[206] S. Hamad, C. R. A. Catlow, S. M. Woodley, S. Lago, J. A. Mejas, J. Phys. Chem. B. **2005**, 109, 15741

[207] S. Li, D. A. Dixon J. Phys. Chem. A **2008**, 112, 6646

[208] S. A. Shevlin, S. M. Woodley, J. Phys. Chem. C. **2010**, 114, 17333

[209] Z.-W. Qu, G.-J. Kroes, J. Phys. Chem. B. **2006**, 110, 8998

[210] Z.-W. Qu, G.-J. Kroes, J. Phys. Chem. C. **2007**, 111, 16808

[211] M. Calatayud, L. Maldonado, C. Minot, J. Phys. Chem. C. **2008**, 112, 16087

[212] M. Calatayud, C. Minot, J. Phys. Chem. C. **2009**, 113, 12186

[213] L. Chiodo, M. Salazar, A. H. Romero, S. Laricchia, F. D. Sala, A. Rubio J. Chem. Phys **2011**, 135, 244704-1

[214] D. J. Mowbray, J. I. Martinez, J. M. Garca-Lastra, K. S. Thygesen, and K. W. Jacobsen. J.Phys. Chem. C **2009**, 113, 12301

[215] T. H. Rana, P. Kumar, A. K. Solanki, R. Skomski, A. Kashyap J. Appl. Phys **2013**, 113, 17B526

Chapter 2

Structure and Spectroscopy of Water-Fluoride Microclusters: A Combined Genetic Algorithm and DFT-Based Study

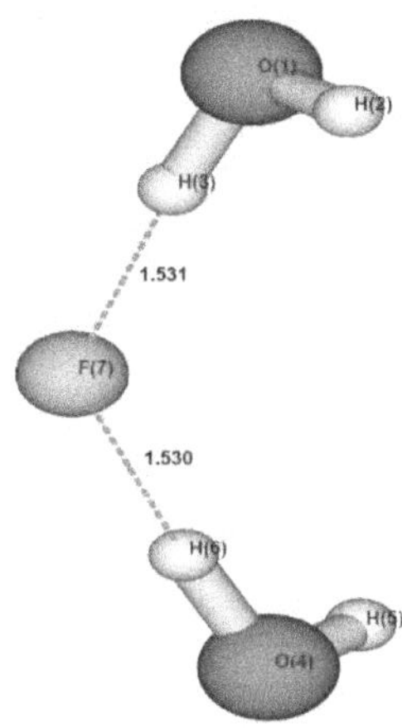

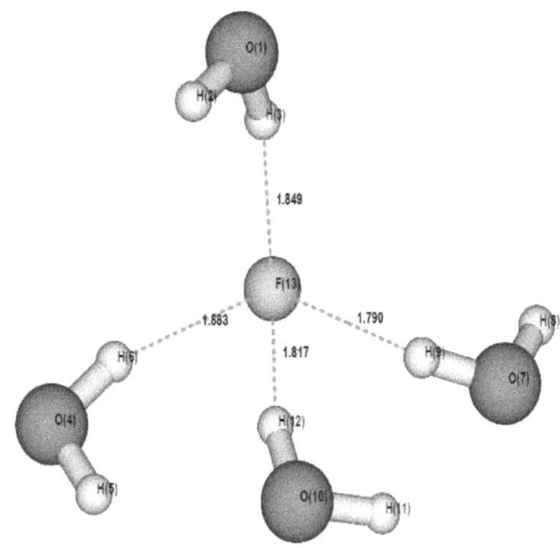

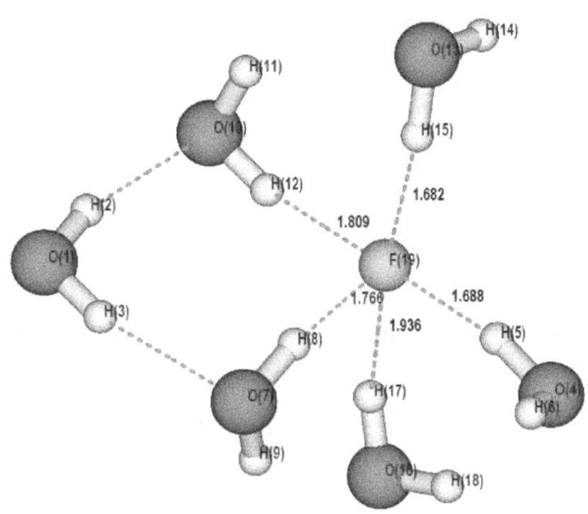

S.G. Neogi, P.Chaudhury J. Comput. Chem. 2012, 33, 629

2.1 Introduction

Cluster Science has become an important area of research for both experimentalists and theoreticians in the last two decades. Both atomic and molecular clusters with diverse bonding patterns among the constituents are studied and characterized with increasing effort. Scientists look into these systems from diverse angles.

For studying pure water clusters the TIP3P, TIP4P, SPCE/POL have been extremely successful [1-4]. These potentials essentially treat the system as a combination of Lennard-Jones and coulombic interaction terms with optimized parameters which have been so given so as to reproduce essential structural features and properties similar to what standard quantum chemical methods give. A more interesting system is when a size-specific solvent cluster like water encapsulates or interacts with a single solute ion like halides or alkali metal ions. In this field the experimental work of Johnson et al.[5,6] deserves special attention though there are numerous other groups who have also done high quality research. The area of interest is how the coordination of water molecules change with size for a given halide ion or how the structure and hence the IR spectrum change with the change of the halide ion for a cluster of a given size. It is expected that halide ions which are smaller in size will prefer encapsulated structure while those which are larger will prefer surface capped ones.

Theoreticians who work with model empirical potentials to study these systems face another stiff challenge. This is the challenge of getting deeper minimum. These potential energy surfaces are extremely rugged and the search for good structures can be a problem for any deterministic optimization methods. Hence there has been an increasing tendency to use truly global and stochastic optimization. These methods borrow their working principle from conventional natural or physical process and construct a mathematical prescription keeping the essence of the natural or physical process. GA[7-10] is one such search algorithm. Ever since its introduction by John Holland [11], it has found increasing use across disciplines ranging from engineering, mathematics, physics and chemistry. GA draws its working philosophy from the genetic process of selection, crossover and mutation. As in conventional genetics these processes in combination have resulted in creating well adapted species under evolution. A mathematical equivalent of selection, crossover and mutation have shown promise as a truly global and stochastic search algorithm, which is able to find out low energy structures of complex large molecules and clusters. GA has been used to solve a host of problems ranging from structure elucidation of atomic and molecular clusters; solution of Schrodinger equation, predicting geometries of polymeric materials etc. The use of GA in elucidating structures of clusters has a rich and long history. Cluster systems ranging from Lennard Jones clusters, Morse clusters, metallic clusters modelled by Gupta potential, molecular clusters like water clusters have received serious attention and work of several leading research groups needs mention. The work of Deaven et al.[12], Pullan et al.[13-15], Niesse and Mayne[16-18], Johnston et al.[19-20], Michaelian et al.[21-24], Chaudhury et al.[25] and Hartke et al.[26-29] deserves special attention. There has been a persistence effort from both experimentalists and theoreticians to explain and understand the structural and spectroscopic properties of water cluster encapsulating a solute ion, both cationic and anionic. Specifically in $(H_2O)_nX^-$ systems experimental works of Bailey et al.[30], Okumura et al.[31,32], Lisy et al.[33] and Johnson et al.[27,28,34] are benchmark works for subsequent workers in this field.

The present chapter deals with GA+DFT method for studying $(H_2O)_nF^-$ cluster systems with n=1-6 and once the structures are obtained the IR spectroscopic modes are calculated using DFT with B3LYP being the functional used and 6-311++G(d,p) being the basis set used. All DFT calculations are performed using GAUSSIAN 03 program.

2.2 The Methodology

2.2.1 The potential for the system

To model the $(H_2O)_nF^-$ system, we have used the interaction potential as given in the paper of Xantheas and Dang[35]. The model uses a rigid three site interaction model potential for water. As is the case for all pair wise potentials, certain fixed charges on the atoms making up the clusters are assigned and also certain values of the Lennard Jones parameters for the atoms. The total energy for the cluster system can be expressed as:

$$U_{tot} = \sum_i \sum_j \left(4\varepsilon \left[\left(\frac{\sigma_{ij}}{r_{ij}} \right)^{12} - \left(\frac{\sigma_{ij}}{r_{ij}} \right)^{6} \right] + \frac{q_i q_j}{r_{ij}} \right) \tag{2.1}$$

Here, r_{ij} is the distance between atoms *i* and *j*, *q* is the respective atomic charge, and the Lennard Jones parameters are σ and ε. The values of all parameters used in the model are given in Table 2.1.

Table 2.1. Parameters used in potential energy function

Atom type	σ(Å)	ε(Kcal/mole)	Charge(q)
O	3.205	0.160	-0.7300
H	0.000	0.000	+0.3650
F	3.358	0.100	-1.0000

2.2.2 Computational Details

Having defined the potential energy of interaction between the constituents of the clusters the goal now would be to design an appropriate search strategy, which will hopefully be able to find out critical points supported by the potential energy function and hence the structures of water-fluoride clusters. Searching out critical points on rugged-multidimensional surface can be a duanting task for any search algorithm. We in the present chapter have used Genetic Algorithm (GA) as the search algorithm. This is a stochastic scheme, based on the biological processes of crossover and mutation and the

Darwinian concept of "*Survival of the Fittest.*" The details of Genetic Algorithm have been discussed in Chapter 1. The GA used by us has a population size of 20. The crossover used was a single point one with the probability of crossover being 50%, that is, 50% of the strings were picked up randomly in a given generation for crossover. Mutation is obviously an infrequent event and the mutation probability was kept at 1% or roughly 1% of the individual variables in the solution pool were picked up to mutate. Our GA based search has yielded reasonable structures for $(H_2O)_nF^-$ systems (n=1-6). Once the GA structures are obtained we proceed to do a DFT calculation with 6-311G++(d,p) being the basis set used and hybrid functional B3LYP using the GA optimized structures as input. Geometries were fully optimized in gas phase and vibrational frequency analyses were performed to confirm that all stationary points represent true minima on the potential energy surface. All DFT calculations for $(H_2O)_nF^-$ system were performed by means of Gaussian 03 suite of ab initio programs[36].

2.3 Results and Discussion

2.3.1 Structures of $(H_2O)_nF^-$ clusters

Our GA based search has yielded reasonable structures for $(H_2O)_nF^-$ systems (n=1-6). The structures are all low energy minima on the potential energy surface described by equation (2.1). All the structures have been generated from a randomly selected initial set of trial solutions. Once the GA evaluated structures were obtained we performed a DFT calculation [with basis set 6-311++G(d,p)] to get the frequencies of the IR modes. The structures returned by DFT calculation closely resembled those obtained by the GA. We give an account of the structural features of the optimized clusters, after the GA stage and the subsequent DFT calculation.

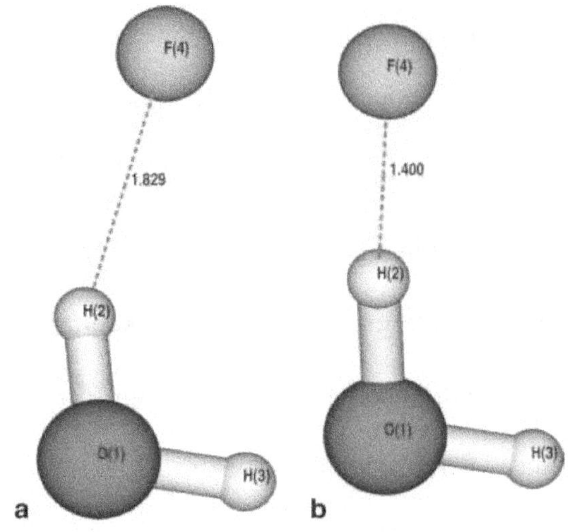

Figure 1.(a) Structure of $(H_2O)_1F^-$ (from GA) (b) Structure of $(H_2O)_1F^-$ (from DFT)

1.$(H_2O)F^-$ The optimized GA and DFT structures, both show proximity of one hydrogen bond with the coordinating F^- ion. The two structures are shown in Fig. 1a and Fig 1b.

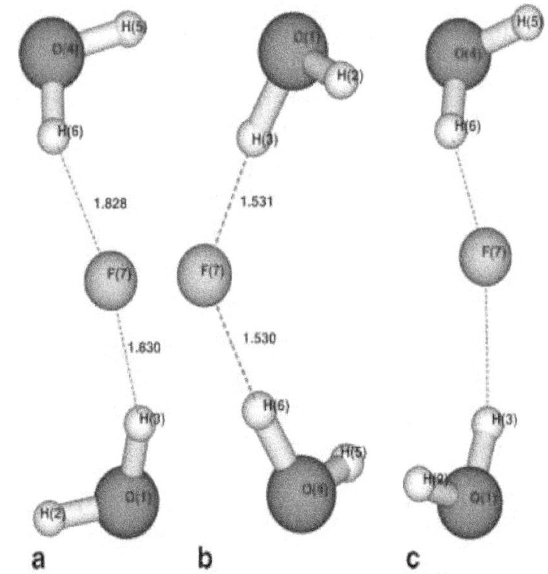

Figure 2.(a) Structure of $(H_2O)_2F^-$ (from GA) (b) Structure of $(H_2O)_2F^-$ (from DFT) (c) local min. structure of $(H_2O)_2F^-$ (from GA)

2.$(H_2O)_2F^-$ For the F^- ion forming cluster with two water molecules both GA and DFT based structures look very similar. In both two hydrogen atoms from two different water molecules are involved in close proximity interaction with the central F^- ion. The other two hydrogen atoms are placed in somewhat anti fashion with respect to each other. This is expected as this would reduce the coulombic interaction between the two electropositive H atoms. The structures are shown in Figure 2a and 2b. Figure 2c shows the local structure from GA.

3.$(H_2O)_3F^-$ In both GA and DFT structures, the arrangement is some-kind of a triangular structure with three hydrogen atoms from three different water molecules being involved in close proximity interaction with the central F^- ion. In both the other three hydrogen atoms are disposed in space in a manner so that the distance between them can be made as large as possible so as to minimize the destabilizing interaction between positively charged constituents. The two structures are shown in Figures 3a and 3b. The higher energy structure generated is shown in Figure 3c. It is significantly different with only hydrogen's from two water molecules coordinating with the F^{-} ion with the appearance of an O—H---O interaction not present in lower energy structure Figure 3a.

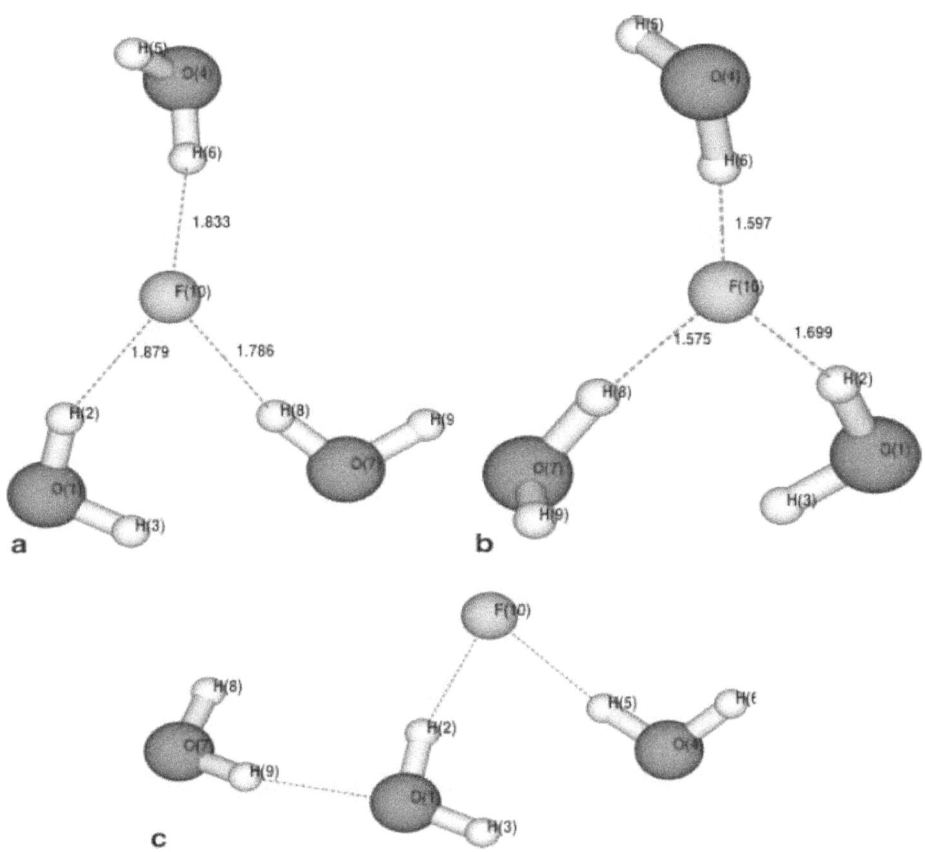

Figure 3. (a) Structure of $(H_2O)_3F^-$ (from GA) (b) Structure of $(H_2O)_3F^-$ (from DFT) (c) local min. structure of $(H_2O)_3F^-$(from GA)

4.$(H_2O)_4F^-$ The two structures from GA and DFT based calculations are shown in Figures 4(a) and4(b). Both structures are essentially similar with minor differences in angular dispositions of the other hydrogen atoms in each water molecule which are not oriented towards the F^- ion. In both the structures four hydrogen atoms are involved in close proximity stabilizing interaction with the central F^- ion. The non-coordinating hydrogen atoms are arranged in such a manner with respect to each other that the destabilizing interaction between similar charged centres is minimized. To again test if our method is

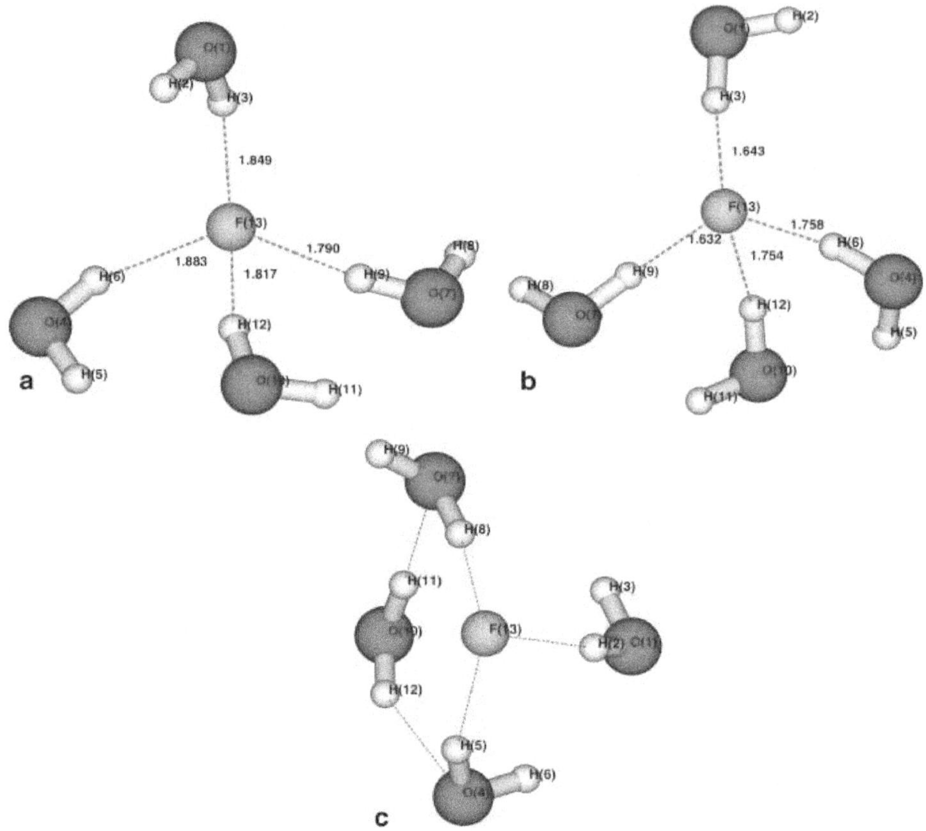

Figure 4. (a) Structure of $(H_2O)_4F^-$ (from GA) (b) Structure of $(H_2O)_4F^-$ (from DFT) (c) local min. structure of $(H_2O)_4F^-$ (from GA)

amenable to find out local minimum structures by repeating the search with a higher (lower bound U_L) than what was used to find out the global minimum, a local structure with lesser number of ionic hydrogen bonds and appearance of O—H---O interaction was found out. The structure is shown in Figure 4(c).

5.$(H_2O)_5F^-$ The structures from the GA and DFT based calculation are shown in Figures 5(a) and 5(b). In both the maximum number of close proximity interactions possible for the systems, i.e.; between the hydrogen atoms in different water molecules and the central F^- ion (five in number) is present. Other than minor differences in spatial orientations, the structures look similar. As is the pattern with the other small size clusters, the non-coordinating hydrogen atoms are rotated in space in such a manner so as to minimize the destabilizing coulombic interactions. Figure 5(c) shows a local energy structure found out using a strategy already discussed.

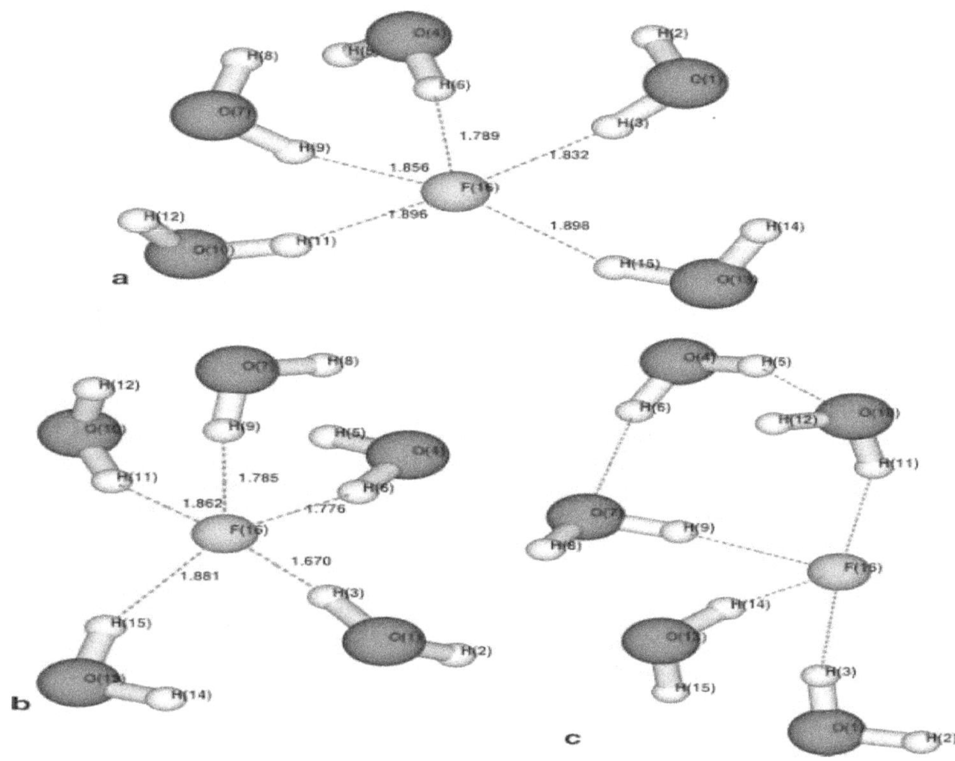

Figure 5. (a) Structure of $(H_2O)_5F^-$ (from GA) (b) Structure of $(H_2O)_5F^-$ (from DFT) (c) local min. structure of $(H_2O)_5F^-$ (from GA)

It has four ionic hydrogen bonds and two O—H---O interactions. The number of hydrogen bonded interactions is less than that present in the global structure [Fig 5(a)].

6.$(H_2O)_6F^-$ The GA and DFT based structures for this system is shown in Figure 6(a) and 6(b). Here we see a marked difference in structural feature as compared to the clusters of smaller size. The maximum number of close proximity O—H---F^- ion interactions is five and not six. The sixth water molecule prefers to have interaction with two other water molecules and is considerably removed in space from the F^- ion. Three water molecules form a cage like structure involving hydrogen bonding interactions. The appearance of a low energy structure showing five O—H---F^- close proximity interaction and not six might be a result of the small size of F^- ion being unable to gather all the six water molecules around itself at close proximity. Figure 6(c) is again a representative local energy structure obtained by us. It has lesser number of ionic hydrogen bonded interactions as compared to the lowest energy structure we have found out [Figure 6(a)].

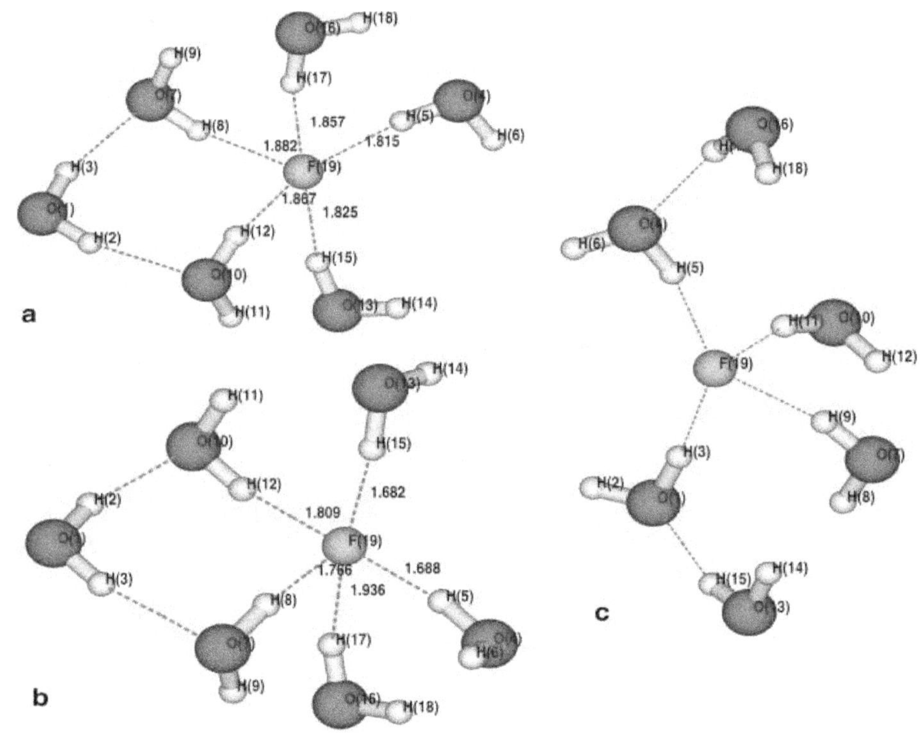

Figure 6. (a) Structure of $(H_2O)_6F^-$ (from GA) (b) Structure of $(H_2O)_6F^-$ (from DFT) (c) local min. structure of $(H_2O)_6F^-$ (from GA)

The structural parameters for the clusters obtained from (GA+DFT) study is shown in Table 2.2. The energy values of the structures obtained from GA and DFT study are reported in Table 2.3. As expected the structures of water-fluoride micro clusters obtained by GA is in close resemblance to the structures obtained by Baik et al. involving pure quantum chemical calculations [37].

Table 2.2 Table showing $\angle H-O-H$ values obtained from this work and literature

Species	Value from this work	Value from literature
$(H_2O)_1F^-$	102.18°	101.8°
$(H_2O)_2F^-$	102.26°	101.9°
$(H_2O)_3F^-$	101.98°	102.2°
$(H_2O)_4F^-$	101.98°	101.4°
$(H_2O)_5F^-$	101.44°	
$(H_2O)_6F^-$	102.3°	

Table 2.3 Energy values of GA and DFT calculation

Species	Global GA (Kcal/mole)	Local GA (Kcal/mole)	Global DFT (Kcal/mole)	Local DFT (Kcal/mole)
$(H_2O)_2F^-$	-32.06	-31.98	-158688.48	-158680.63
$(H_2O)_3F^-$	-48.23	-42.59	-206679.68	-206669.76
$(H_2O)_4F^-$	-62.51	-58.43	-254677.15	-254667.11
$(H_2O)_5F^-$	-75.73	-72.34	-302668.35	-302659.82
$(H_2O)_6F^-$	-86.70	-77.59	-350659.55	-350644.49

To check the correspondence of the structural parameters (bond lengths and bond angles) to results already published, we report the bond angles (H-O-H) of the different clusters obtained by us (after GA+DFT calculation) in Table 2.2, along with those already reported in literature [38]. The results after only the GA stage of the calculation are in Figure 1a, 2a, 3a, 4a, 5a and 6a. The various interatomic lengths obtained by us after GA+DFT based search are shown in Figure 1b, 2b, 3b, 4b, 5b and 6b. The values are close to those already reported in the literature [37]. We have performed DFT calculations with the local minimum structures found out by GA, to check whether the energy obtained by doing a DFT calculation is indeed higher than the DFT energies of the corresponding global minimum structures. The results are presented in Table 2.3. The energies given in Table 2.3 after the calculation with GA and DFT are quite different in magnitude. This is because the energies after GA calculation represents the potential energy of the system as defined specifically by the empirical potential energy surface while obviously the DFT energy represents the total electronic energy after a calculation at quantum chemical level.

2.3.2 Spectroscopy of $(H_2O)_nF^-$ clusters

Once the GA evaluated structures are obtained we performed a DFT calculation [with basis set 6-311++G(d,p)] and functional B3LYP to get the frequencies of the IR modes. We give an analysis of the spectroscopic features of the optimized clusters after the DFT calculation. The spectral details are shown in Table 2.4. One consistency in the entire set of results is that wherever a hydrogen bonding is seen to occur involving O—H----F^-, there is a discernable red shift in O—H stretching compared to the stretching frequency of an O—H entity not involved in hydrogen bonding.

Table 2.4 Spectral data of $(H_2O)_nF^-$ clusters

O—H stretching frequencies (cm^{-1})		
Species	Calculated	Experimental
$(H_2O)_1F^-$	2231, 3876	1523, 3687
$(H_2O)_2F^-$	2818,2982,3877.23, 3877.66	2520, 3700
$(H_2O)_3F^-$	3045, 3171, 3388, 3838, 3879, 3881	2888, 3127, 3386, 3706
$(H_2O)_4F^-$	3222, 3305, 3454, 3526, 3820, 3831, 3880, 3882	3069, 3359, 3700
$(H_2O)_5F^-$	3331, 3476, 3516, 3591, 3639, 3777, 3803, 3813, 3881	3140, 3386, 3542, 3698
$(H_2O)_6F^-$	3350, 3393, 3510, 3579, 3663, 3668, 3716, 3768, 3803, 3811, 3875, 3884	

For $(H_2O)F^-$, whereas the peak at 3876cm^{-1}[O(1)—H(3)] does not involve any hydrogen bonding, a peak occurs at 2231cm^{-1} involving O(1) and H(2) [Figure(1b)]. For $(H_2O)_2F^-$ the stretching at 3877.23cm^{-1} and 3877.66cm^{-1} are for O—H stretches not taking part in H-bonding, the major contribution being from bond stretches O(4)—H(5) and O(1)—H(2). The hydrogen bonded stretches at 2817cm^{-1} and 2982cm^{-1} involves O(1)—H(3) and O(4)—H(6) bonds which are directed towards F^- ion [Figure(2b)]. The $(H_2O)_3F^-$ cluster shows similar features. Peaks at 3045cm^{-1}, 3171cm^{-1} and 3388cm^{-1} are due to the stretches O(7)—H(8), O(4)—H(6) and O(1)—H(2), all being directed towards the F^-. The stretches at 3879cm^{-1}, 3838cm^{-1} and 3881.2cm^{-1} involve O(4)—H(5), O(7)—H(9) and O(1)—H(3) which are directed away from F^- [Figure (3b)]. For $(H_2O)_4F^-$ the red shifted peaks arising from H-bonded stretches are the 3222cm^{-1}, 3305cm^{-1}, 3454cm^{-1} and 3526cm^{-1} with contributions from O(7)—H(9), O(1)—H(3), O(10)—H(12), O(4)—H(6) bond stretching. All of these are H-bonded. The stretches at 3820 cm^{-1}, 3831 cm^{-1}, 3880 cm^{-1} and 3881.96 cm^{-1} are for bonds not involved in H-bonding [Figure (4b)]. In $(H_2O)_5F^-$ results in the similar lines are observed. The H-bonded stretches are at 3331cm^{-1}, 3476cm^{-1}, 3516cm^{-1} and 3693cm^{-1} involving bonds O(1)—H(3), O(4)—H(6), O(7)—H(9), O(10)—H(11) and O(13)—H(15). The peaks at 3777 cm^{-1}, 3802.11 cm^{-1}, 3802.99 cm^{-1}, 3813 cm^{-1} and 3888 cm^{-1} have contributions from nonhydrogen bonded stretches [Figure (5b)]. In $(H_2O)_6F^-$ the peaks at 3350cm^{-1}, 3393cm^{-1}, 3510cm^{-1} and 3579$cm^{-1}$are due to hydrogen bonded stretches of the type O—H----F^- with the specific bond stretches making the contribution being O(4)—H(5), O(13)—H(15), O(7)—H(8) and O(10)—H(12). The peaks at 3663cm^{-1} and 3668cm^{-1} are due to O—H---O type with the bonds contributing being O(16)—H(17),O(1)—H(2) and O(1)—H(3). The other stretches at 3716cm^{-1}, 3768cm^{-1}, 3803cm^{-1}, 3811cm^{-1}, 3875cm^{-1} and 3884cm^{-1} are due to non-hydrogen bonded types [Figure (6b)]. The experimental O—H stretching frequencies for these systems are also included in Table 2.4. These results are due to the works of the groups of Johnson [28] and Lisy [39]. In all these systems, the frequencies obtained by us are blue shifted with respect to the experimental frequencies. This is because the mode of calculation used in deciphering the frequencies is based on fully harmonic approximation. The discrepancies are due to the real anharmonic effects being present which the theory used presently do not incorporate. Excepting the $(H_2O)F^-$ and $(H_2O)_2F^-$, where very large red shifts in the ionic hydrogen bond is present, for the other clusters a scaling factor of about 0.95 can provide the matching of the major peaks, namely the free O—H stretch and the ionic hydrogen bonded stretch, in the theoretical calculation with the experimental spectra.

2.4 Conclusion

We have shown that a strategy based on a combined GA and DFT can be a viable tool to study structure and spectroscopy of water-fluoride micro clusters. This simple strategy of studying structure and spectroscopy supported by an empirical potential can answer many questions related to spectral shift associated with size change of clusters.

2.5 References

[1] N. L. Jorgensen, J. Chandrasekhar, J. D. Madura, R. W. Impey, M. L. Kelin, J. Chem. Phys. **1983**, 79, 926

[2] W. L. Jorgensen, J. Am. Chem. Soc. **1981**,103, 335

[3] D. J. Wales, M. P. Hodges, Chem. Phys. Lett. **1998**, 286, 65

[4] P. Chaudhury, R. Saha, S. P. Bhattacharya, Chem. Phys. **2001**, 270, 277

[5] M. A. Johnson, M. L. Alexander, W. C. Lineberger, Chem. Phys. Lett. **1984**, 112, 285

[6] J. R. Roscioli, E. G. Dicken, M. A. Johnson, J. Phys. Chem. A **2006**, 110,4943

[7] D. M. Deaven, K. M. Ho, Phys. Rev. Lett **1995**, 75, 288

[8] P. Chaudhury, S. P. Bhattacharya Chem. Phys. **1999**, 241, 213

[9] R. Saha, P. Chaudhury, S. P. Bhattacharya, Phys. Lett. A **2001**, 291, 397

[10] C. Roberts, R. L. Johnston, N. T. Wilson, Theor. Chem. Acc. **2000**, 104, 123

[11] J. Holland, Adaptation in Natural and Artificial Systems; MIT Press, Cambridge, Massachussets, **1992**

[12] D. M. Deaven, J. R. Morris, K. M. Ho, N. Tit, Chem. Phys. Lett. **1996**, 256, 195

[13] W. J. Pullan J. Chem. Inf. Modell. **1997**, 37, 1189

[14] W. J. Pullan Comp. Phys. Comm. **1997**, 107, 137

[15] W. J. Pullan J. Comput. Chem. **1997**, 18, 1096

[16] J. A. Niesse, H. R. Mayne, Chem. Phys. Lett. **1996**, 261, 576

[17] J. A. Niesse, H. R. Mayne, J. Chem. Phys. **1996**, 105, 4700

[18] J. A. Niesse, H. R. Mayne, J. Comput. Chem. **1997**, 18, 1233

[19] R. L. Johnston, Dalton Trans. **2003**, 4193

[20] R. Ismail, R. L. Johnston, Phys. Chem. Chem. Phys. **2010**, 12, 8607

[21] K. Michaelian, Chem. Phys. Lett. **1998**, 293, 202

[22] I. L. Garzon, N. Rendon, K. Michaelian, Phys. Rev. B **1999**, 60, 2000

[23] K. Michaelian, Am. J. Phys. **1998**, 66, 231

[24] K. Michaelian, I. L. Garzon, M. R. Beltran, P. Ordejon, D. Portal-Sanchez, J. M. Soler, E. Artacho, Phys. Rev. Lett. **1998**, 81, 1600

[26] B. Hartke, Chem. Phys. Lett. **1995**, 240, 560

[27] B. Hartke, Chem. Phys. Lett. **1996**, 258, 144

[28] S. K. Gregurick, M. H. Alexander, B. Hartke, J. Chem. Phys. **1996**, 104, 2684

[29] B. Hartke, M. Schutz, H. J. Werner, Chem. Phys. **1998**, 239, 561

[30] C. G. Bailey, J. Kim, M. A. Johnson, J. Phys. Chem. **1996**, 100, 16782

[31] M. S. Johnson, K. T. Kuwata, C. K. Wong, M. Okumura, Chem. Phys. Lett. **1996**, 260, 551

[32] J. H. Choi, K. T. Kuwata, Y. B. Cao, M. Okumura, J. Phys. Chem. A **1998**, 102, 503

[33] O. M. Cabarcos, C. J. Weinheimer, J. M. Lisy, S. S. Xantheas, J. Chem. Phys **1999**, 110, 5

[34] P. Ayotte, S. B. Nielsen, G. H. Weddle, M. A. Johnson, J. Phys. Chem. A **1999**, 103, 10665

[35] S. S. Xantheas, L. X. Dang, J. Phys. Chem **1996**, 100, 3989

[36] M. J. Frisch, Gaussian 03, Revision E.01, Gaussian, Inc, Wallingford CT, **2004**

[37] J. Baik, H. M. Lee, S. B. Suh, D. Majumder, K. S. Kim, J. Chem. Phys. **1999**, 110, 9116

[38] J. Kim, H. M. Lee, S. B. Suh, D. Majumdar, K. S. Kim, J. Chem. Phys. **2000**, 113, 5259

[39] O. M. Cabarcos, C. J. Weinheimer, J. M. Lisy, S. S. Xantheas, J. Chem. Phys. **1999**, 110, 5

Chapter 3

Structure and Spectroscopic Aspects of Water-Halide Ion Clusters: A study Based on Conjunction of Stochastic and Quantum Chemical Methods

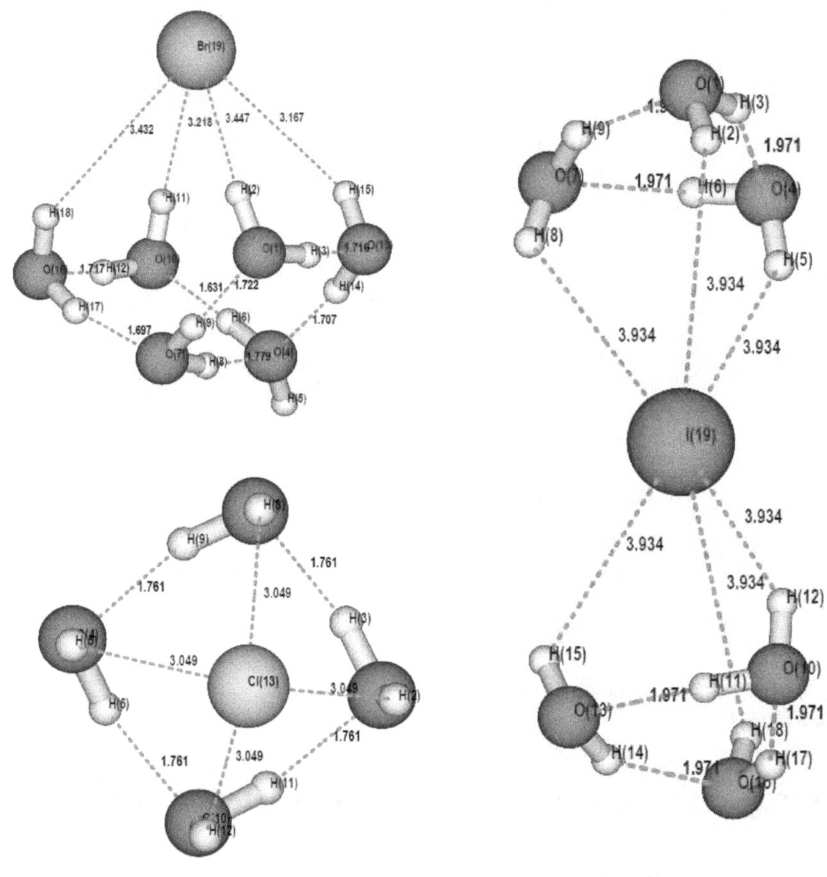

S. G. Neogi, P. Chaudhury J. Comput. Chem. 2013, 34, 471

3.1 Introduction

The study of atomic and molecular clusters is an actively pursued area of research in contemporary chemical physics. It is a challenging field to be engaged in as it throws up new and interesting features for experimentalists to find out and for theoreticians to provide logical explanations consistent with experimental findings. The area of hydrated halide ion clusters is an important system for study from the point of view of bonding, structure and spectroscopy. Experimentally the creation of size specific hydrated halide ion clusters has become a reality, and once such clusters are produced, one can go on to measure its spectroscopic properties. Several groups across the world have contributed extensively towards the study of these systems [1-42]. The work of Johnson et al. deserves a special mention in this context, especially the detail study of a solute halide ion interacting and forming cluster with a single solute molecule. A simple system like $(H_2O)_nX^-$(X^-=halide ion) can show up spectroscopic details which can be a challenge for any theoretical model or method to explain in the right perspective[13, 15]. The structural properties of these systems can become extremely complicated, with the appearance of many local minima other than the usually sought after global one. The structures especially when the numbers of solvated water molecules are increased, show two types of hydrogen bonding, the O—H -----X^- type and O—H----O type. These hydrogen bonds also leave their mark on the IR spectroscopy of these systems. The usual pattern is the appearance of high intensity red shifted peaks, compared to the position of the O—H stretch in free water. The magnitude of the red shift observed varies with the nature of the central halide ion, as well as the size of the system. These red-shifted peaks, especially the ones formed from ionic hydrogen bond (O—H---X^-), is a typical signature of these systems as compared to the IR spectroscopy of $(H_2O)_nY^+$(Y^+=alkali metal ion) where the ionic hydrogen bonds are absent[6,7,43-48]. There have been various approaches adopted, ranging from calculations using empirically constructed potential energy surfaces (PES) or using quantum chemical methods at various levels of theory [49-116]. High level quantum chemical calculations can be costly and so to circumvent these difficulties, well calibrated empirical potentials have been used by several workers, and these are mostly modifications of TIP3P, TIP4P, SPCE/POL models for studying pure water clusters[117-120]. All these potentials are a combination of Lennard-Jones, Coulombic interactions and interactions arising out of contribution from polarizability factors. These include both pair-wise additive interaction terms as well as terms involving more than two body effects. Recently there has been an impressive application in $(H_2O)_nX^-$ systems[121] where the central ion is a molecular halide ion. The strategy of using empirical potential energy surfaces, though elegant and computationally quick may not be that easy to use in practice. These surfaces support a large number of minima and the task of finding the lowest one from the myriad of possibilities is a non-trival task. Optimization is the key step in finding out a stable structure. Conventional gradient based optimization techniques are found to be inadequate in most cases as the structure which is generally found out is the one which is achieved by the quickest descent in gradient norm from the initial starting point. Hence the guarantee that a global structure is reached is entirely fortuitous and one has to scan the entire search space starting from different initial guess geometries to arrive at the correct result. To overcome this difficulty the trend of using global optimizers are on the rise. Global optimizers in general use principles of random walk and stochasticity to arrive at good quality solutions quickly. The explicit dependence of gradient to move along the surface is done away with as this can lead to being trapped in high energy local minimum. Genetic Algorithm (GA) is one such technique which uses operators, which are mathematical equivalents of Selection, Crossover and Mutation processes in conventional genetics. The details of GA are discussed in chapter 1 of the thesis.

In the present chapter we find out both global and local minima structures in $(H_2O)_nX^-$ systems (X=Cl,Br,I) using GA to search a well parameterized empirical potential energy surface. Once the GA optimized structures are obtained we intend to do a DFT level calculation with the GA optimized structure as the input. We intend to check the correspondence between the structures obtained by GA and DFT techniques. We also theoretically calculate the various O—H stretching frequencies and compare with the experimental works reported in literature. Moreover we evaluate vertical detachment energies and NPA (natural population analysis) charge for all the cluster systems.

3.2. The Methodology

A. The Potential for the system

To describe the $(H_2O)_nX^-$ system , we have used the interaction potential as suggested by Xantheas and Dang [74]. In this model a rigid three site model interaction potential for water is assumed. As with all pairwise potentials, there are fixed atomic charges and Lennard-Jones potential parameters defined for each atomic site. The total energy can thus be expressed as;

$$U_{total} = \sum_i \sum_j \left(4\varepsilon \left[\left(\frac{\sigma_{ij}}{r_{ij}} \right)^{12} - \left(\frac{\sigma_{ij}}{r_{ij}} \right)^{6} \right] + \frac{q_i q_j}{r_{ij}} \right)$$

The model potential is given by sum of Coulomb and LJ interactions. Here, r_{ij} is the distance between atoms i and j, q is the respective atomic charge, and the Lennard Jones parameters are σ and ε. The values of all parameters used in the model are given in Table 1. and Table 2. Having defined the potential energy of interaction between the constituents of the clusters, the objective now would be to construct an appropriate search strategy which will be successful in finding out the critical points in the potential energy function of $(H_2O)_nX^-$ system. We have used Genetic Algorithm as the search strategy. The details of Genetic Algorithm are described in Chapter 1 of the thesis. The GA used by us has a population size 20. The crossover used was single point. Mutation probability was kept at 1% roughly.

Table 1. Parameters used in potential energy function for Cl^- and Br^-			
Atom type	σ(Å)	ε(Kcal/mole)	Charge(q)
O	3.1536	0.1550	-1.040
H	0.0000	0.0000	+0.520
Cl	4.3210	0.1000	-1.000
Br	4.6030	0.1000	-1.000

Table 2. Parameters used in potential energy function for I^-			
Atom type	σ(Å)	ε(Kcal/mole)	Charge(q)
O	3.2340	0.1825	-1.0380
H	0.0000	0.0000	+0.5190
I	5.1245	0.1000	-1.0000

B. Quantum Chemical Calculations

Our contention is that the structures of $(H_2O)_nX^-$ systems (X = Cl,Br,I) obtained by a GA search on an empirical potential energy surface are of high quality and are close to the ones obtained by doing a quantum chemical calculation at the DFT level of theory. Once the GA structures are obtained, we proceed to do a DFT calculation with 6-311++ G(d,p) being the basis set used and functional B3LYP, using the GA optimized structures as input. We have used Stuttgart pseudo potential for calculations of $(H_2O)_nI^-$ clusters. As the GA structures are quite close to the minima (local as well as global) in DFT calculations, convergence is extremely quick. We also extract the vibrating frequencies of various modes from the quantum chemical calculation. All calculations are performed using GAUSSIAN 09[122] package.

3.3 Results and Discussions

A. Structures of $(H_2O)_nX^-$ clusters

Our GA [123-125] based search has yielded various low-lying energy structures for $(H_2O)_nX^-$ systems (n=1-6). Once the GA evaluated structures were obtained a DFT calculation [with basis set 6-311++G(d,p)] is performed to get the frequencies of the IR

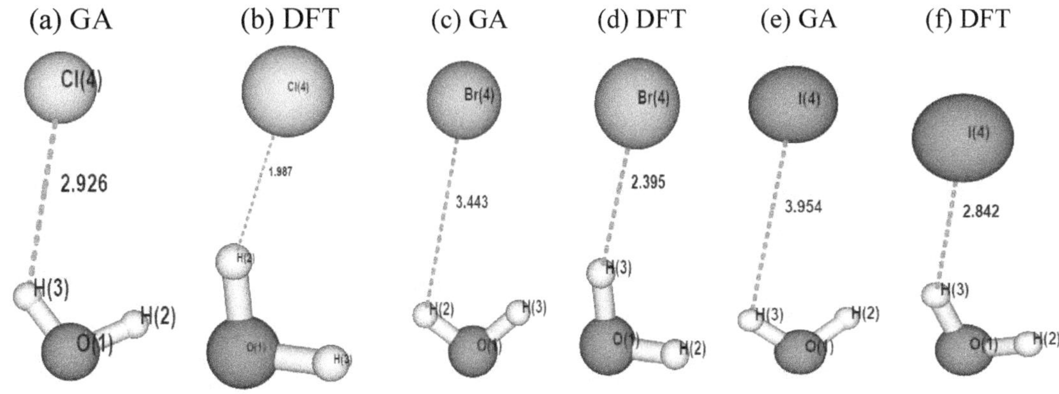

Figure 1. Structure of $(H_2O)Cl^-$, $(H_2O)Br^-$,$(H_2O)I^-$

modes. We give an account of the structural features of the optimized clusters, after the GA stage and the subsequent DFT calculation. H_x represents the hydrogen bonded to halide anion, H_w represents the hydrogen bonded to other water and H_f represents free non-bonded hydrogen. The structures of $(H_2O)_nF^-$ systems (n=1-6)[126] are fundamentally different from the clusters with the larger halide ions. This is due to the difference in size of the ions. While F^- prefers more encapsulated structures, but the higher halide ions show their preference for surface capped ones. The optimized GA and DFT structures, both show a proximity of one hydrogen with the coordinating X^- ion. The optimized structures for $(H_2O)X^-$ [X=Cl,Br,I] are shown in Figure 1.

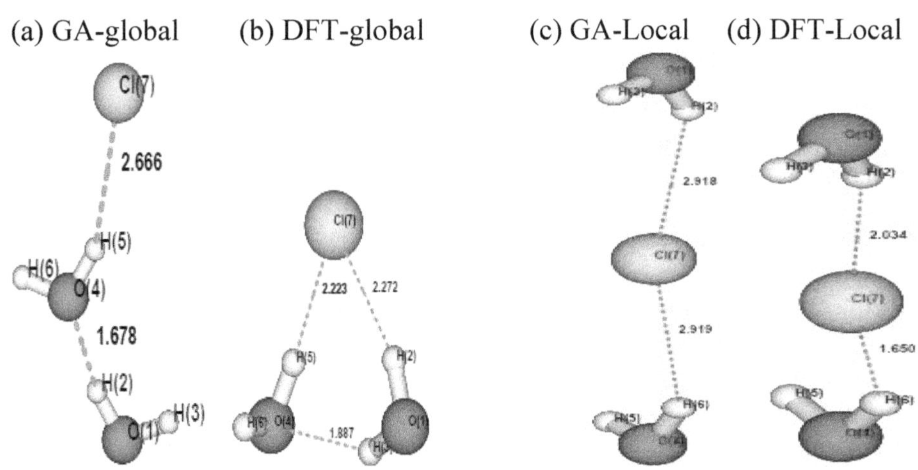

Figure 2. Structure for $(H_2O)_2Cl^-$

(a) GA-Global (b) DFT-Global (c) GA-Local (d) DFT-Local

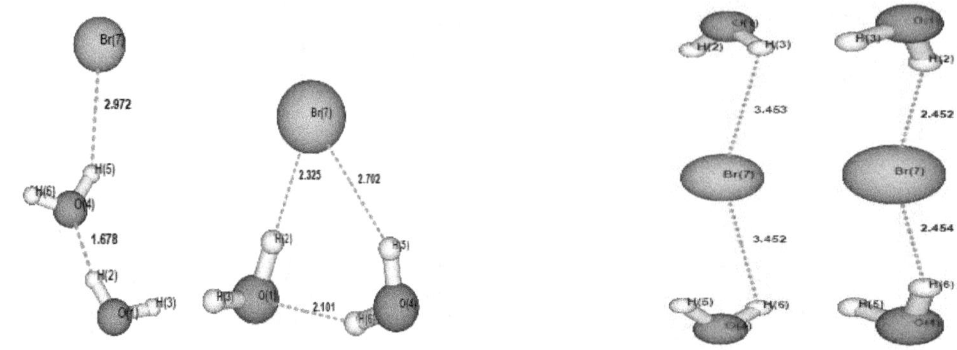

Figure 3. Structure for $(H_2O)_2Br^-$

(a) GA-Global (b) DFT-Global (c) GA-Local (d) DFT-Local

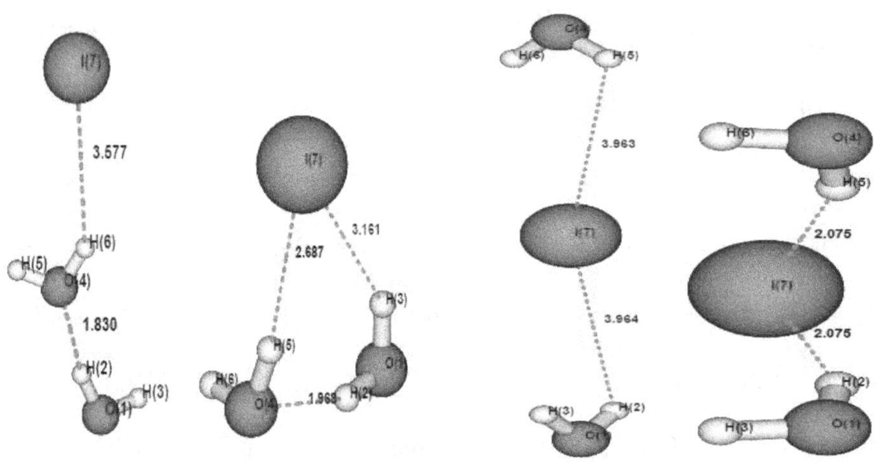

Figure 4. Structure for $(H_2O)_2I^-$

As the electronegativity of the halogen atom decreases and the atomic radius increases from Cl to I, the X----H_x strength decreases which is reflected in the structures showing greater X----H_x length as one goes from $(H_2O)Cl^-$ to $(H_2O)I^-$. The structures of $(H_2O)_2X^-$ clusters are shown in Figures 2, 3 and 4 respectively. In case of $(H_2O)_2Cl^-$ the GA based structure show one O—H bond in close proximity towards the co-ordinating Cl^- ion, involved in a O—H weak interaction. The other water molecule is attached with a O—H----O hydrogen bond interaction to the ionic hydrogen bonded water molecule. In the DFT structure the two O—H----Cl^- interactions are not of equal strength as one of them is in closer proximity than the other. For $(H_2O)_2Br^-$ and $(H_2O)_2I^-$ both the GA and DFT structures show the same structural pattern as that obtained for $(H_2O)_2Cl^-$.The structures of $(H_2O)_3X^-$ clusters are shown in Figures 5, 6 and 7 respectively. The global minimum of $(H_2O)_3X^-$ cluster has C_3 symmetry with all three water monomers occupying equivalent positions. The three water monomers form the base of a trigonal pyramid with the halide anion at its apex. The structure is formed by a water trimer interacting with the halide ion and is the first representative structure of a surface complex. In both GA and DFT evaluated structures, the arrangement is some kind of pyramidal structure with three hydrogen atoms from three different water molecules being involved in close proximity interaction with the central X^-. On going from $(H_2O)_3Cl^-$ to $(H_2O)_3I^-$ the inter water hydrogen bonds gets somewhat stronger at the expense of X---H_x weakening of three water molecules.

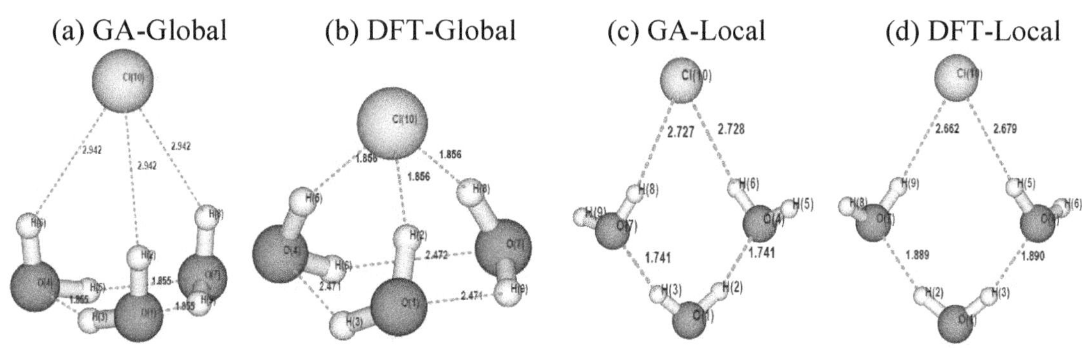

Figure 5. Structure for $(H_2O)_3Cl^-$

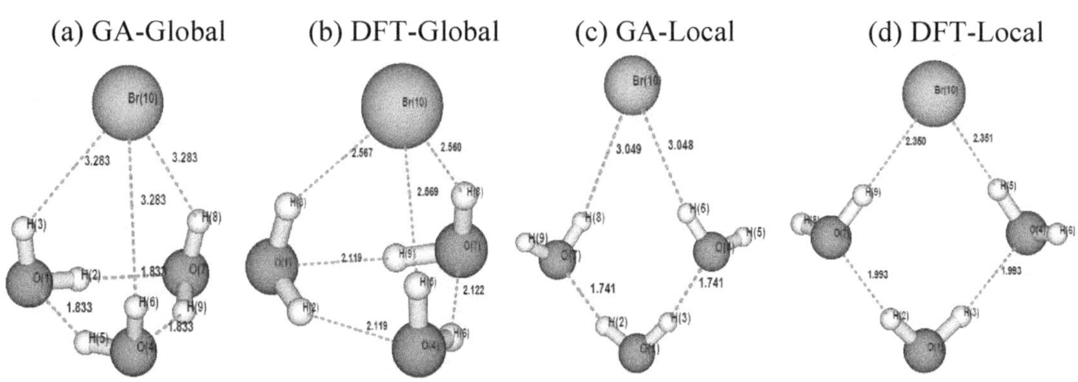

Figure 6. Structure of $H_2O)_3Br^-$

The structures of $(H_2O)_4X^-$ clusters are shown in Figures 8, 9, and 10. For $(H_2O)_4X^-$ clusters the global minimum has C_4 symmetry, where the halide sits at the apex of pyramid with the four water molecules arranged so that each has two hydrogen bonds with the

(a) GA-Global (b) DFT-Global (c) GA-Local (d) DFT-Local

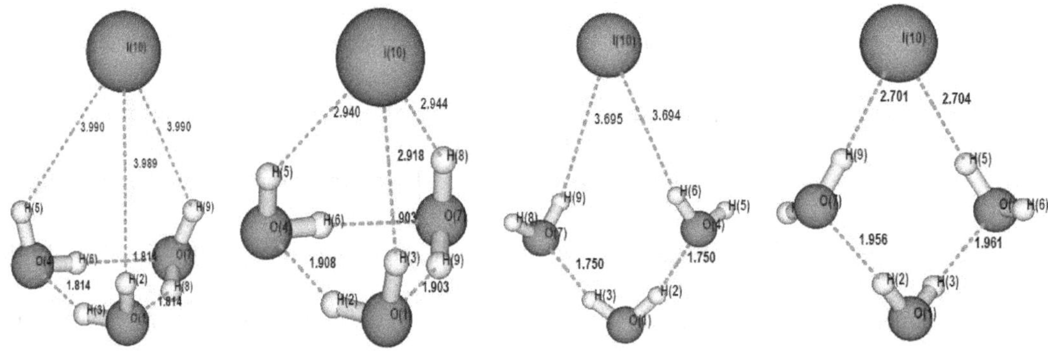

Figure 7. Structure of $(H_2O)_3I^-$

neighbouring waters and one hydrogen bond to halide. The structure corresponds to a well-defined four body water cluster that interacts with the halide anion. All the four water molecules engage in O—H---O hydrogen bonded interactions (four in number) giving the structure its stability.

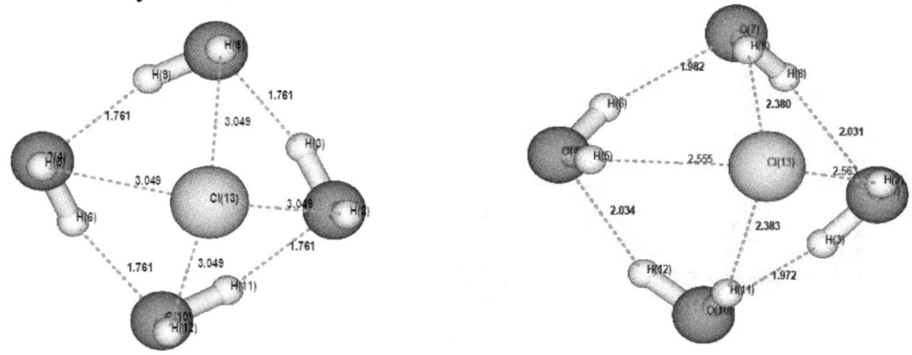

(a) GA-Global (b) DFT-Global

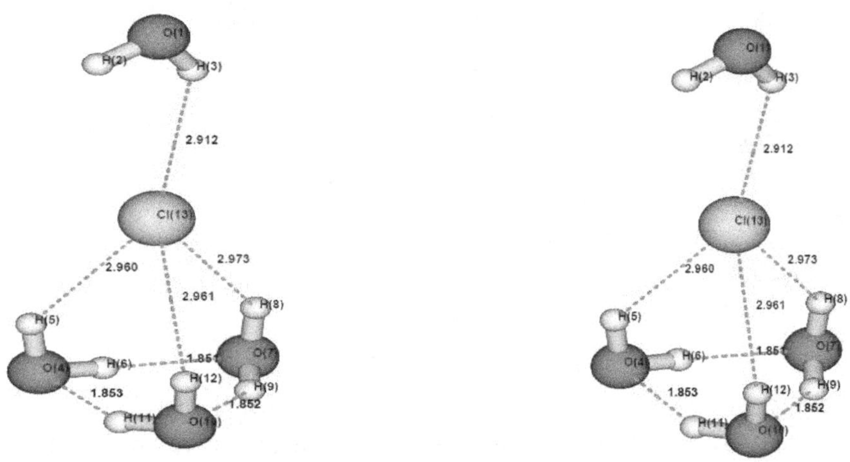

(c) GA-Local (d) DFT-Local

Figure 8. Structure of $(H_2O)_4Cl^-$

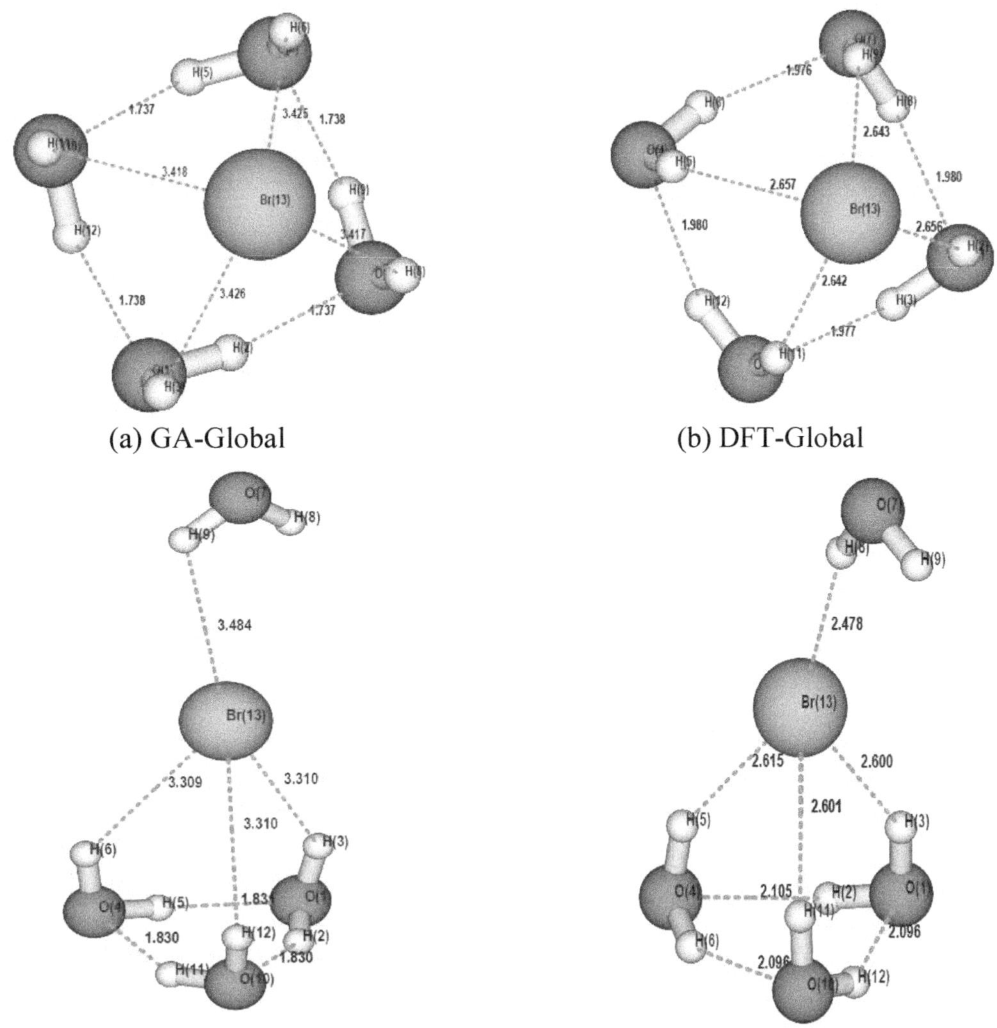

(a) GA-Global (b) DFT-Global

(c) GA-Local (d) DFT-Local

Figure 9. Structure of $(H_2O)_4Br^-$

The structures of $(H_2O)_5X^-$ are shown in Figures 11, 12 and 13. The GA and DFT structures are broadly similar with four water molecules being disposed nearer to the co-ordinating X^- ion and one being away from it. The water molecule disposed away from the central X^- ion is involved in three O—H---O hydrogen bonded interaction. The water molecules are arranged in space in such a manner so as to minimize the destabilizing interaction between similar charged centres forming a cage network involving O—H----O hydrogen bonded interactions. The distance of hydrogen atom bonded to the central halide anion increases as one goes from $(H_2O)_5Cl^-$ to $(H_2O)_5I^-$ with consequent increase in inter-water hydrogen bond strength.

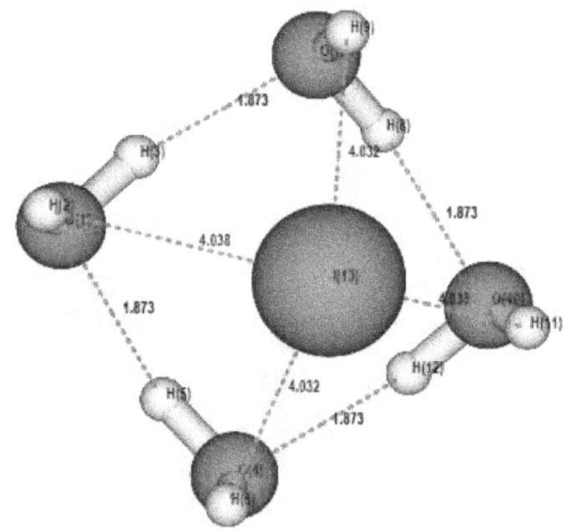

(a) GA-Global

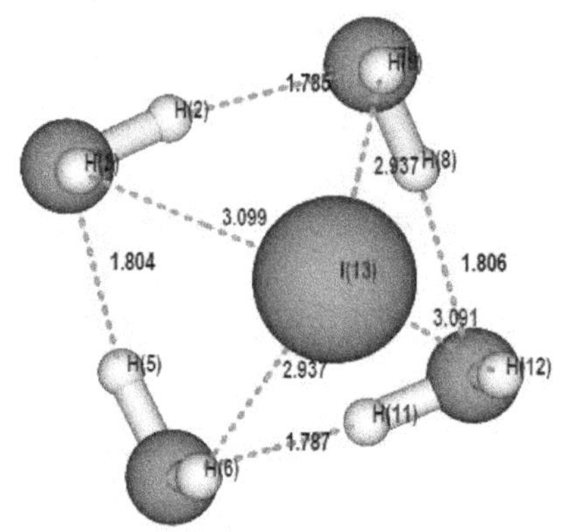

(b) DFT-Global

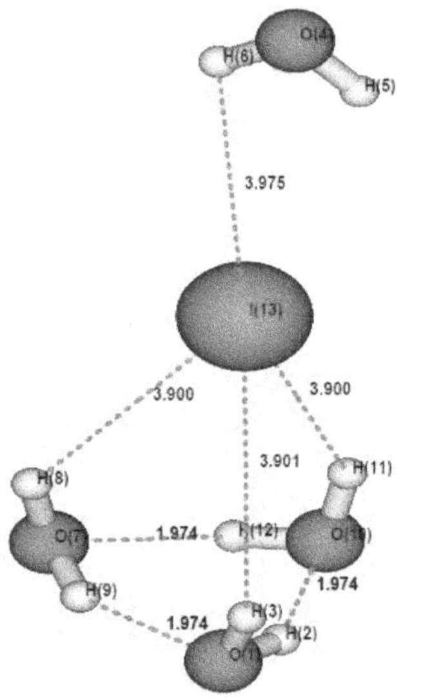

(c) GA-Local

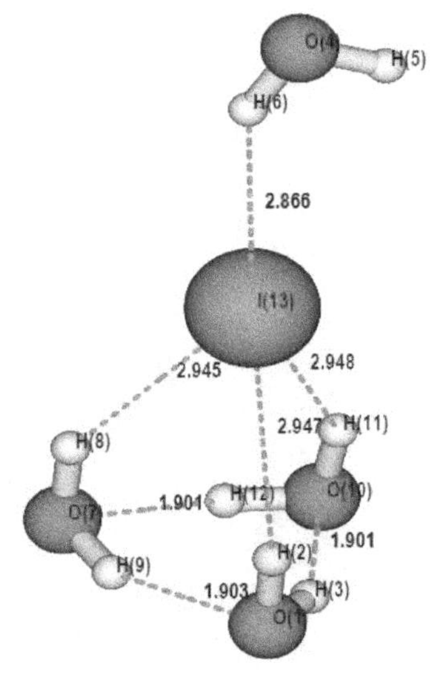

(d) DFT-Local

Figure 10. Structure of $(H_2O)_4I^-$

The structures of $(H_2O)_6X^-$ clusters are shown in Figures 14, 15 and 16. The $(H_2O)_6X^-$ complex possesses four water molecules in the primary hydration shell and two water molecules in the secondary hydration shell respectively. The O—H group that is not involved in hydrogen bonding is also present. The closer H_2O molecules engage in O—H---X^- weakly bonded interaction, while the other two H_2O molecules can form O—H---O hydrogen bonded interaction. The GA and DFT based structures are more or less similar. The overall structure is a cradle shaped structure involving four O—H---X^- close proximity interaction as well as seven O—H---O hydrogen bonded interaction imparting stability to the structure.

Table 3. Energy values of GA and DFT calculation of $(H_2O)_nCl^-$ Clusters

Size(n)	Global GA (Kcal/mole)	Local GA (Kcal/mole)	Relative Shift (Kcal/mole)	Global DFT (Kcal/mole)	Local DFT (Kcal/mole)	Relative Shift (Kcal/mole)
2	-28.9466	-25.2561	3.6905	-384825.3512	-384824.0165	1.3350
3	-47.1436	-45.9756	1.1680	-432817.9318	-432817.2183	0.7135
4	-66.5378	-58.5000	8.0378	-480809.4674	-480806.6737	2.7937
5	-81.2830	-77.5061	3.7769	-528800.5861	-528797.6551	2.9310
6	-103.6272	-90.2937	13.3335	-576791.5011	-576785.9784	5.5227

Table 4. Energy values of GA and DFT calculation of $(H_2O)_nBr^-$ Clusters

Size(n)	Global GA (Kcal/mole)	Local GA (Kcal/mole)	Relative Shift (Kcal/mole)	Global DFT (Kcal/mole)	Local DFT (Kcal/mole)	Relative Shift (Kcal/mole)
2	-26.6045	-21.9441	4.6604	-1711315.7030	-1711313.7240	1.9790
3	-43.2957	-42.0408	1.2549	-1759307.6780	-1759306.6240	1.0540
4	-62.5278	-53.3299	9.1979	-1807298.9210	-1807294.8490	4.0720
5	-76.7157	-72.3093	4.4064	-1855289.0810	-1855285.5850	3.4960
6	-99.0101	-83.8312	15.1789	-1903280.1390	-1903273.1150	7.0240

Table 5. Energy values of GA and DFT calculation of $(H_2O)_nI^-$ Clusters

Size(n)	Global GA (Kcal/mole)	Local GA (Kcal/mole)	Relative Shift (Kcal/mole)	Global DFT (Kcal/mole)	Local DFT (Kcal/mole)	Relative Shift (Kcal/mole)
2	-21.5434	-17.3409	4.2025	-103156.8386	-103151.7239	5.1147
3	-38.0316	-36.2678	1.7638	-151130.4271	-151122.7822	7.6449
4	-51.0021	-42.6990	8.3031	-199100.2799	-199091.0575	9.2224
5	-64.1391	-58.9578	5.1813	-247066.3598	-247060.7834	5.5764
6	-79.5394	-67.6025	11.9369	-295037.6081	-295028.9917	8.6164

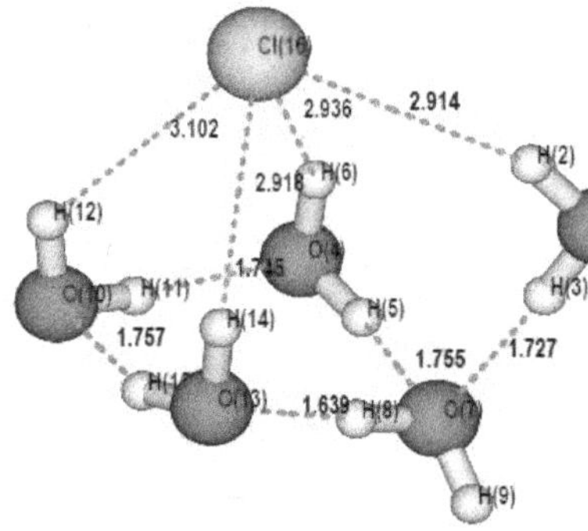

(a) GA-Global

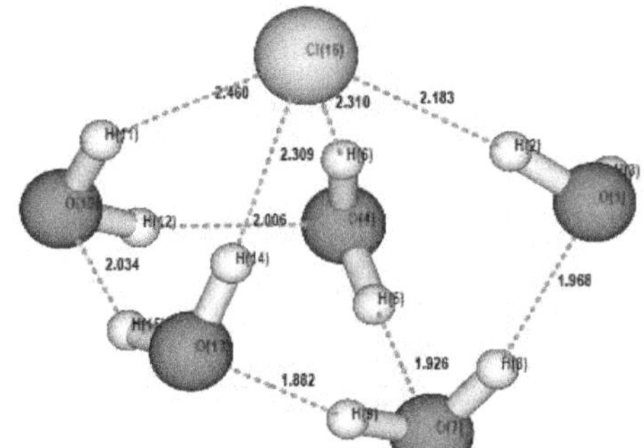

(b) DFT-Global

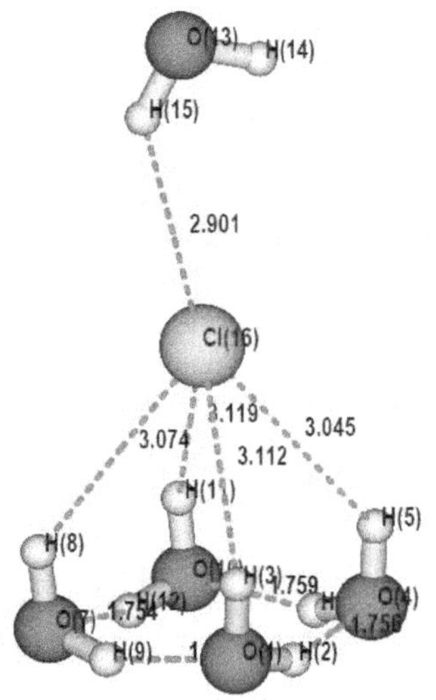

(c) GA-Local

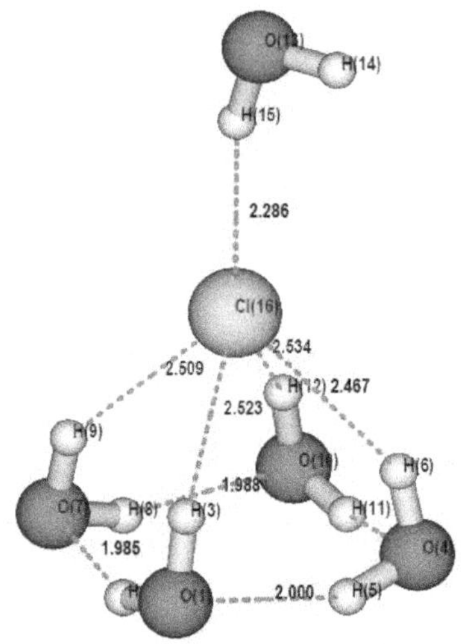

(d) DFT-Local

Figure 11. Structure of $(H_2O)_5Cl^-$

The energies of the structures (global as well as local) obtained from GA and DFT calculations are summarized in Tables 3, 4 and 5. Tables 6, 7 and 8 shows the comparison of bond angle data obtained from our study and those available in literature [57] for the chloride, bromide, and iodide cluster systems, respectively.

Table 6. Table showing bond ∠ H-O-H values of $(H_2O)_n\ Cl^-$ obtained from this work and those reported in literature.

Species	Value from this work	Value from literature
1 $(H_2O)_1Cl^-$	101.1°	100.6°
2 $(H_2O)_2Cl^-$	101.5°	100.4°
3 $(H_2O)_3Cl^-$	104.5°	100.2°
4 $(H_2O)_4Cl^-$	101.1°	99.9°
5 $(H_2O)_5Cl^-$	101.1°	
6 $(H_2O)_6Cl^-$	101.8°	

Table 7. Table showing bond ∠ H-O-H values of $(H_2O)_n\ Br^-$ obtained from this work and those reported in literature.

Species	Value from this work	Value from literature
1 $(H_2O)_1Br^-$	101.2°	99.7°
2 $(H_2O)_2Br^-$	101.6°	100.4°
3 $(H_2O)_3Br^-$	101.5°	100.5°
4 $(H_2O)_4Br^-$	101.3°	100.2°
5 $(H_2O)_5Br^-$	102.6°	
6 $(H_2O)_6Br^-$	103.3°	

Table 8. Table showing bond ∠ H-O-H values of $(H_2O)_n\ I^-$ obtained from this work and those reported in literature.

Species	Value from this work	Value from literature
1 $(H_2O)_1I^-$	99.7°	99.4°
2 $(H_2O)_2I^-$	100.8°	100.6°
3 $(H_2O)_3I^-$	101.4°	100.9°
4 $(H_2O)_4I^-$	101.4°	100.6°
5 $(H_2O)_5I^-$	102.2°	
6 $(H_2O)_6I^-$	103.0°	

B.Spectral Details of $(H_2O)_nX^-$ clusters

The spectral details are shown in Tables 9, 10 and 11. Compared to an ion free $(H_2O)n$ cluster the gross feature in $(H_2O)_nX^-$ spectra is the gain in intensity of (O—H---X^-) stretch at the expense of (O—H---O) stretch. This is due to strengthening of ionic hydrogen bonds and is marked by a dramatic red shift of the O—H stretching mode from 3500cm^{-1} to around 3200cm^{-1} in all the halide-water cluster system. The results which we have presented are consistent with experimental observation. The present discussion on hydrated halides (Cl^-, Br^- and I^-) is important in view of experimental O—H stretching frequencies. In case of $(H_2O)X^-$ cluster the free O—H_f frequency remains almost unchanged with respect to the free water. The hydrogen bonded O—H_x length with respect to X^- ion is elongated considerably and consequently O—H_x stretching frequencies show large redshift with respect to free water.

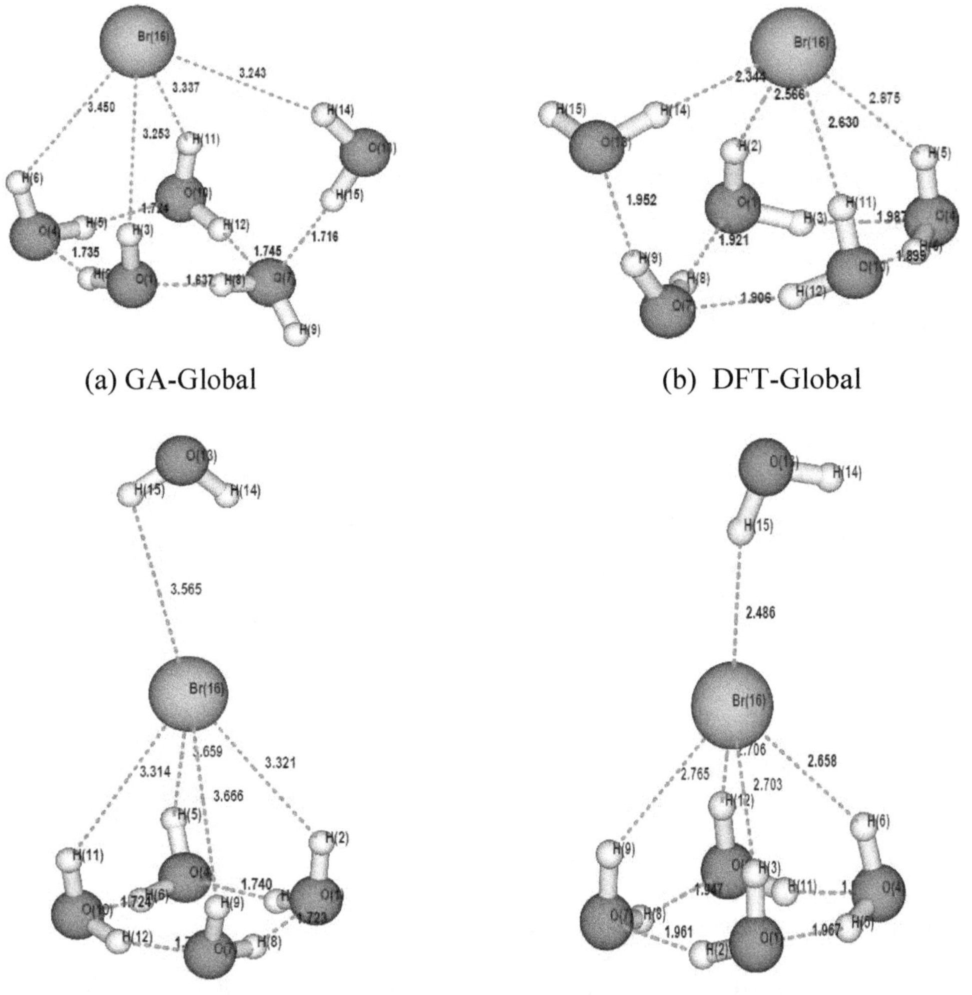

(a) GA-Global (b) DFT-Global

(c) GA-Local (d) DFT-Local

Figure 12. Structure of $(H_2O)_5Br^-$

It is observed that as the strength of ionic hydrogen bond decreases from Cl^- to I^-, the amount of red-shift decreases. The O—H stretching frequencies in halide-water clusters could be classified as (1) nonbonded free [ω(O—H_f)], (2) hydrogen bonded to another water[ω(O—H_w)] and (3) hydrogen bonded to halide atom [ω(O—H_x)]. It has been found that, for elongation of the O—H bonds due to hydrogen bonding the O—H stretching frequencies show red-shift with respect to the average symmetric and asymmetric stretches of water [3870cm^{-1} in DFT/6-311++G(d,p)]. The magnitude of the shift follows the same trend as the elongation of the O—H bond length, i.e. ω(O-H_f)< ω(O-H_w)< ω(O-H_x). The O—H stretching frequencies for the different halide-water clusters are indicative of their

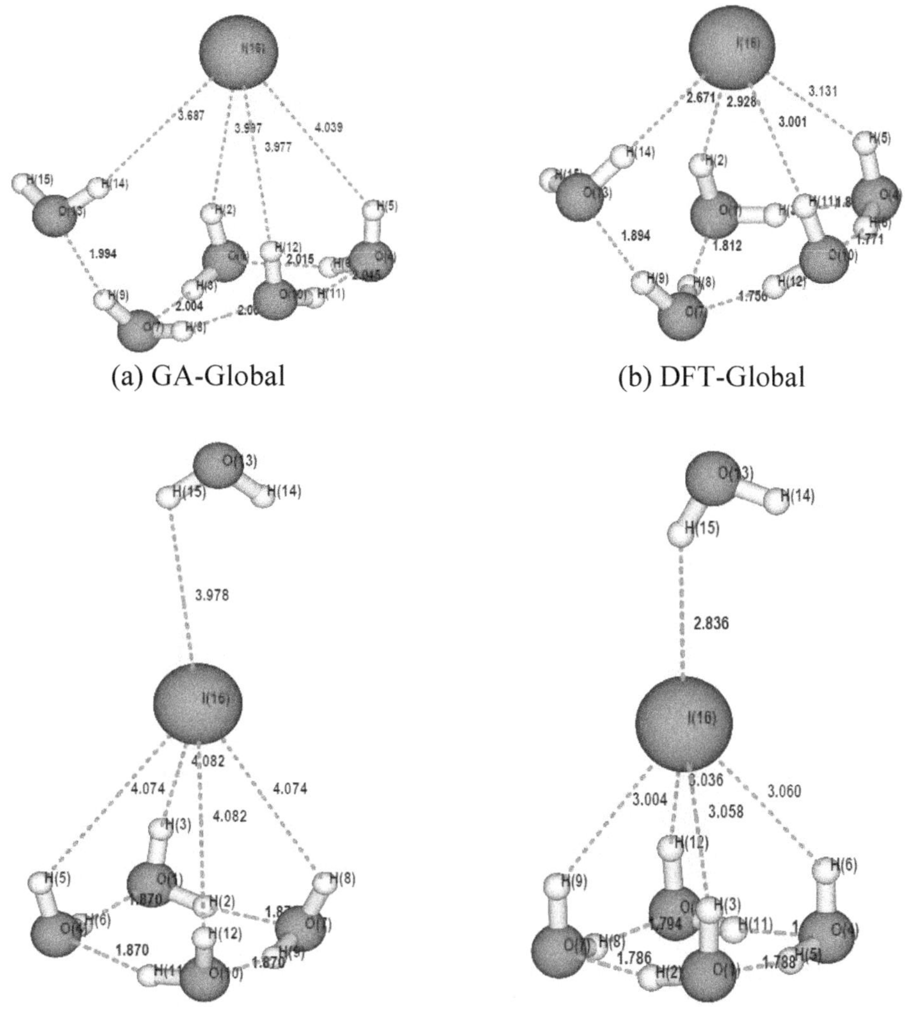

(a) GA-Global (b) DFT-Global

(c) GA-Local (d) DFT-Local

Figure 13. Structure of $(H_2O)_5Br^-$

probable structural arrangement when compared with the experimental frequencies. Since the ω(O—H_x) strength is dependent on the strength of (X---H_x) interaction, it permits to interpret bonding features. The frequency data presented in Tables 9, 10 and 11 indicate that as the X----H_x bond strength increases from I^- to Cl^-, with the increase of electronegativity and decrease of ionic radii of halides, the redshift in O—H_x increases. For any particular anion-water cluster, the experimental spectra show small blue shifts in the positions of ionic-hydrogen bond stretches as the cluster size increases from n=3 to 6 in $(H_2O)_nX^-$. This can be explained on the basis that an increase in the number of the water molecules in the cluster enables the O—H---O bonds to regain strength with slight weakening of O—H---X bonds. Our study of IR-spectra is consistent with the experimental results observed by Okumura [27, 28] and co-workers.

Table 9. Spectral data of $(H_2O)_n$ Cl^- clusters: (O-H)stretching frequencies.

Species	DFT/harmonic	DFT/anharmonic	Experimental
$(H_2O)_1Cl^-$	3337, 3372, 3875	3140, 3324, 3685	3130, 3285 3690, 3698
$(H_2O)_2Cl^-$	3300, 3319, 3401 3639, 3756, 3871	3129, 3290, 3374 3583, 3686	3130, 3375 3633, 3686
$(H_2O)_3Cl^-$	3381, 3382, 3425, 3569 3600, 3713, 3733, 3690	3302, 3322, 3541 3563, 3566, 3734	3310, 3354, 3391 3585, 3600
$(H_2O)_4Cl^-$	3406, 3412, 3414, 3445 3604, 3621, 3626, 3634 3669, 3671, 3675, 3687	3430, 3446, 3303, 3368 3480, 3423, 3451, 3407 3308, 3288, 3260, 3311	3408, 3590 3694
$(H_2O)_5Cl^-$	3407, 3526, 3574, 3596 3626, 3641, 3658, 3666 3736, 3873	3258, 3282, 3345, 3396 3364, 3458, 3455, 3468 3414, 3680	3233, 3441 3548, 3696
$(H_2O)_6Cl^-$	3154, 3312, 3390, 3400 3422, 3437, 3462, 3489 3529, 3539, 3570, 3587 3625, 3653, 3672, 3696 3719, 3888	3210, 3262, 3265, 3285 3303, 3318, 3331, 3334 3345, 3360, 3365, 3414 3430, 3448, 3518, 3551 3771	

Table 10. Spectral data of $(H_2O)_n$ Br^- clusters: (O-H)stretching frequencies.

Species	DFT/harmonic	DFT/anharmonic	Experimental
$(H_2O)_1Br^-$	3325, 3476, 3873	3238, 3682	3270, 3690
$(H_2O)_2Br^-$	3310, 3388, 3392 3684, 3744, 3867	3254, 3287, 3464 3575, 3691	3207, 3438 3625, 3680
$(H_2O)_3Br^-$	3373, 3375, 3416, 3605 3607, 3622, 3698, 3716 3717	3207, 3215, 3338 3340, 3382, 3511 3543, 3548	3244, 3302, 3410 3565, 3584
$(H_2O)_4Br^-$	3412, 3413, 3441, 3574 3614, 3617, 3645, 3664 3674, 3680, 3700	3261, 3269, 3271, 3281 3398, 3406, 3431, 3442 3457, 3461	3245, 3466 3567
$(H_2O)_5Br^-$	3297, 3395, 3417, 3432 3444, 3450, 3503, 3555 3578, 3627, 3650, 3659 3679, 3761, 3873	3256, 3264, 3319, 3333 3376, 3389, 3422, 3435 3456, 3463, 3467, 3488 3706	
$(H_2O)_6Br^-$	3153, 3313, 3378, 3391 3416, 3435, 3457, 3477 3511, 3553, 3575, 3591 3628, 3668, 3678, 3690 3719, 3887	3208, 3241, 3276, 3283 3300, 3320, 3342, 3356 3359, 3374, 3402, 3408 3431, 3464, 3489, 3529 3715	

Table 11. Spectral data of $(H_2O)_n\ I^-$ clusters: (O-H)stretching frequencies.

Species	DFT/harmonic	DFT/anharmonic	Experimental
$(H_2O)_1I^-$	3582, 3698, 3869	3324, 3695	3270, 3690
$(H_2O)_2I^-$	3292, 3381 3471, 3692, 3773, 3863,	3289, 3374 3583, 3686	3331, 3500 3616, 3675
$(H_2O)_3I^-$	3363, 3367, 3412, 3626 3656, 3658, 3700, 3710	3441, 3448, 3452, 3473 3483, 3488	3440, 3462, 3548 3565
$(H_2O)4I^-$	3442, 3521, 3577, 3607 3706, 3714, 3715, 3734	3512, 3441, 3447, 3463 3527	3543, 3683 3733
$(H_2O)_5I^-$	3285, 3393, 3410, 3426 3449, 3467, 3511, 3531 3562, 3602, 3661, 3692 3728 ,3777, 3871	3419, 3480, 3491, 3500 3572, 3685	3496, 3526, 3551 3581, 3621, 3671 3711, 3751
$(H_2O)_6I^-$	3164, 3312, 3371, 3381 3412, 3439, 3446, 3458 3494, 3531, 3578, 3640 3641, 3684, 3711, 3716 3722, 3887	3115, 3185, 3230, 3250 3258, 3270, 3324, 3351 3384, 3420, 3425, 3471 3489, 3500, 3513, 3709 3771, 3229	

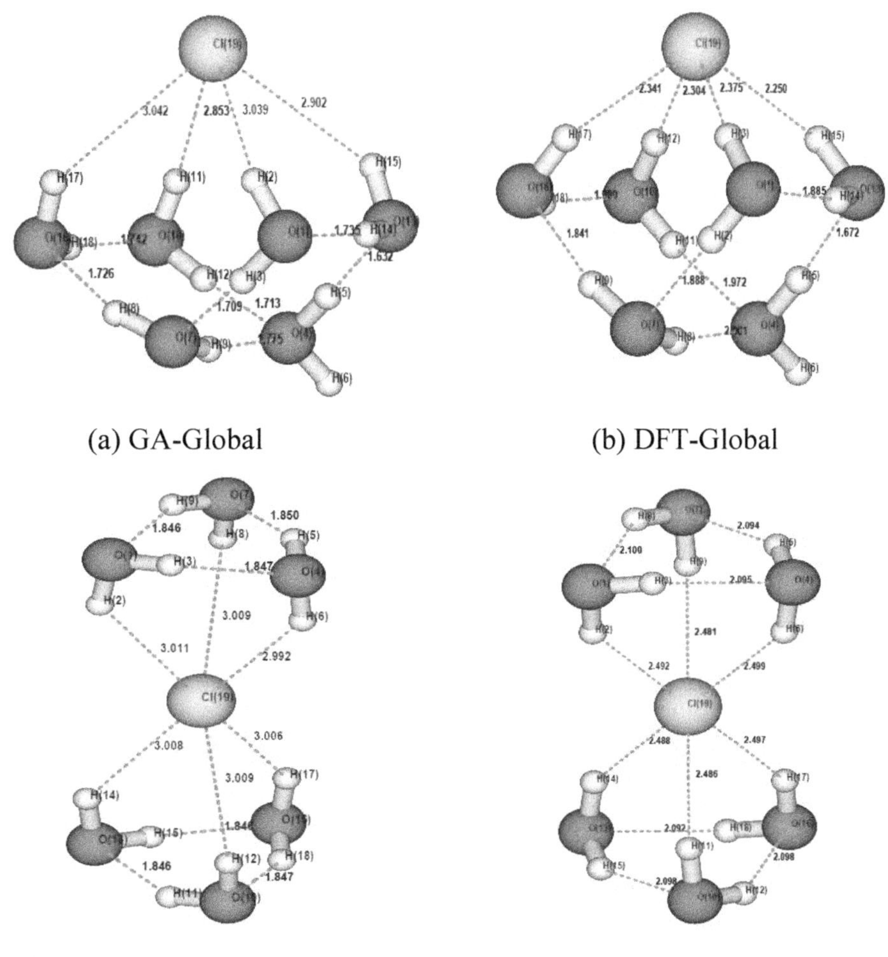

(a) GA-Global (b) DFT-Global

(c) GA-Local (d) DFT-Local

Figure 14. Structure of $(H_2O)_6Cl^-$

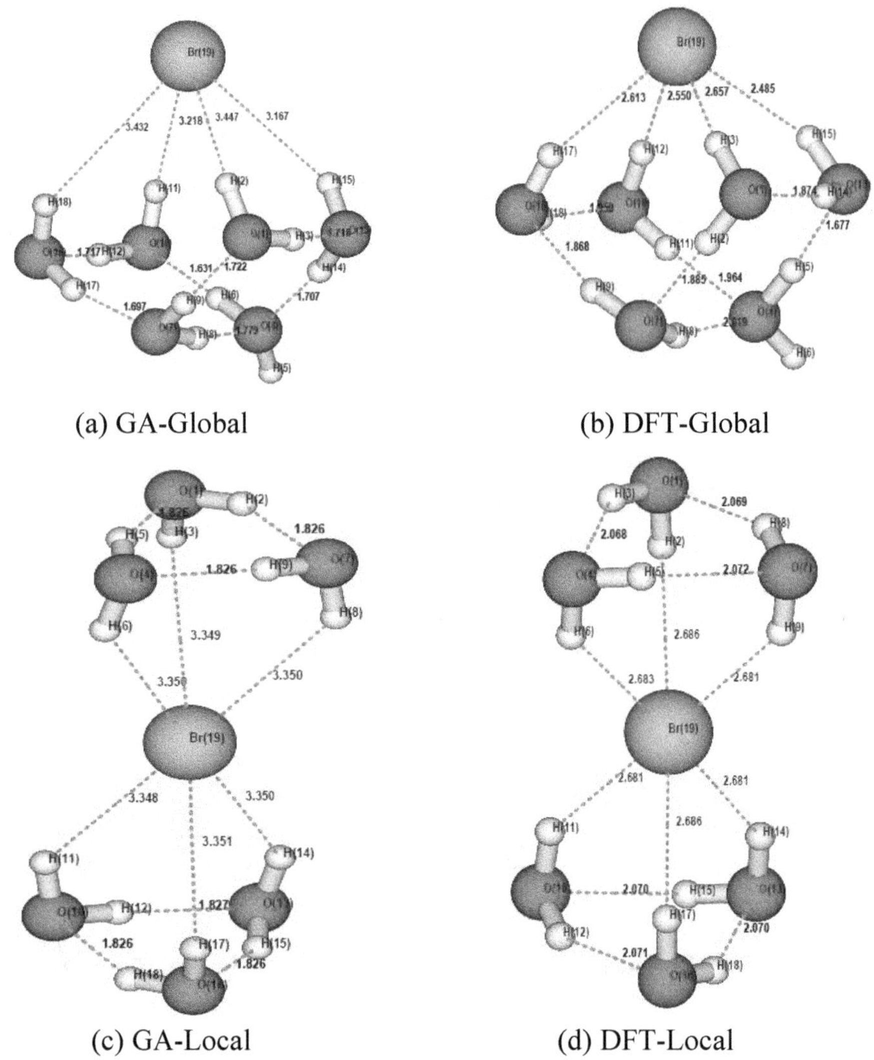

(a) GA-Global (b) DFT-Global

(c) GA-Local (d) DFT-Local

Figure 15. Structure of $(H_2O)_6Br^-$

The blue shift of the ionic H-bonded O—H stretch with the increase of water molecules is attributed to the weakening of the ionic H-bond as more water molecules bind to the ion. The stronger H-bond in $(H_2O)_nCl^-$ system is most blue shifted with the formation of water-water network. The $(H_2O)_nI^-$ and $(H_2O)_nBr^-$ system which already has a weaker bond to a single water molecule, suffers less reduction upon formation of the water sub clusters, leading to a smaller proportional shift. The O—H_x show large red frequency shifts and large IR intensities due to large dipole changes by the vibrational modes. The single proton acceptor or double proton acceptor type of water monomer shows strong interaction with the halide ion and, thus, large frequency shifts.

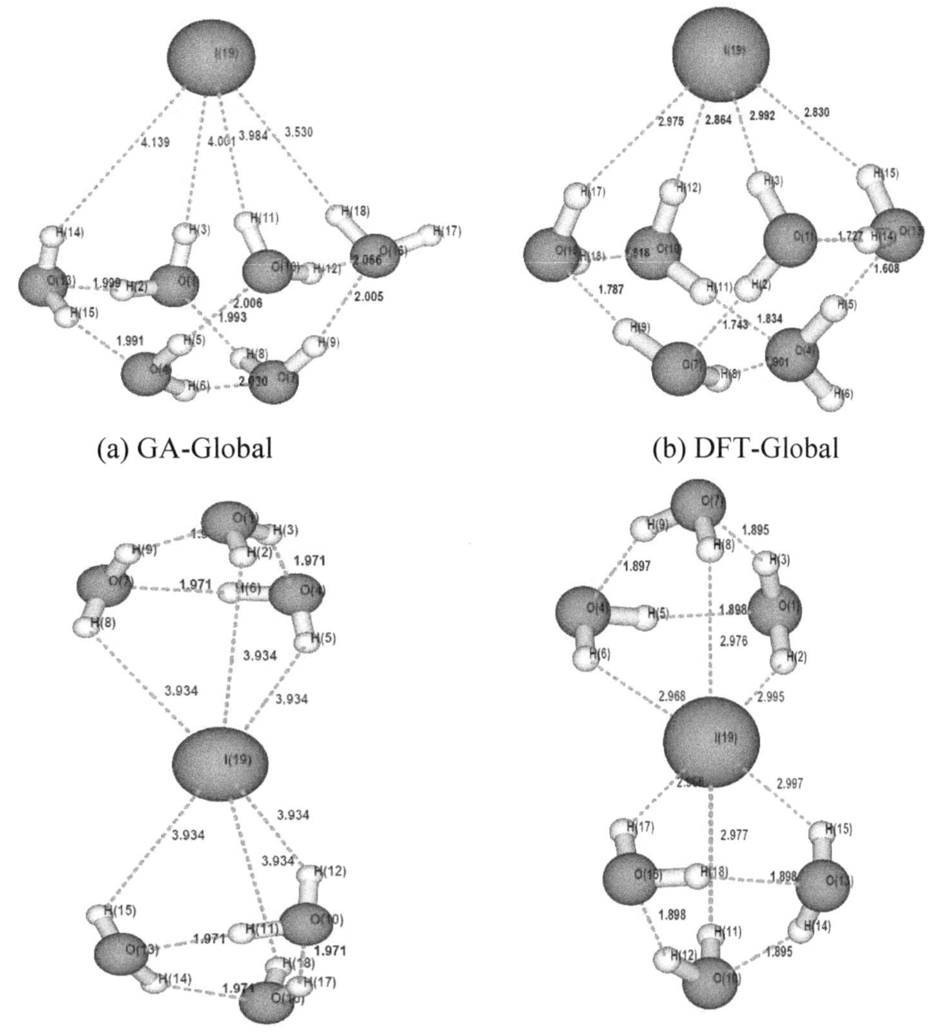

(a) GA-Global (b) DFT-Global

(c) GA-Local (d) DFT-Local

Figure 16. Structure of $(H_2O)_6I^-$

The experimental O—H stretching frequencies for these systems are also included in Tables 9, 10 and 11. These results are due to the works of the groups of Johnson et al.[9, 11, 13], Lisy and co-workers[6,7], Ayotte et al.[19-24], keley et al.[25], Corcelli et al.[26]. In these systems, the frequencies obtained by us are blue shifted with respect to the experimental frequencies. This is because the mode of calculation used in deciphering the frequencies is based on fully harmonic approximation. For $(H_2O)_nX^-$ clusters a scaling factor of about 0.93 can provide the matching of the major peaks, namely the free O—H stretch and the ionic hydrogen bonded stretch, in the theoretical calculation with the experimental spectra. To see whether anharmonic corrections can induce any change in the position of ir-peaks calculated harmonically, we have included anharmonic corrections in our calculations. We find a very good improvement in the spectral peaks. There is a very close correspondence between the experimental frequencies and the anharmonic frequencies obtained for the corresponding water halide clusters. The correspondence of the experimental frequencies to anharmonic frequencies is nearly 99%.

C. VDE of $(H_2O)_nX^-$ clusters

The profile for the change in Vertical Detachment Energy (VDE) with cluster size (n) for $(H_2O)_nX^-$(X=Cl, Br, I) is shown in Figure 17. From the figure, it is evident that for a given size(n), VDE decreases as we move from Cl^- to Br^- and then to I^- that is VDE [$(H_2O)_nCl^- > (H_2O)_nBr^- > (H_2O)_nI^-$]. This is completely in keeping with the expected trend since the negative charge on the halide ion is most stabilised for the $(H_2O)_nCl^-$ cluster.

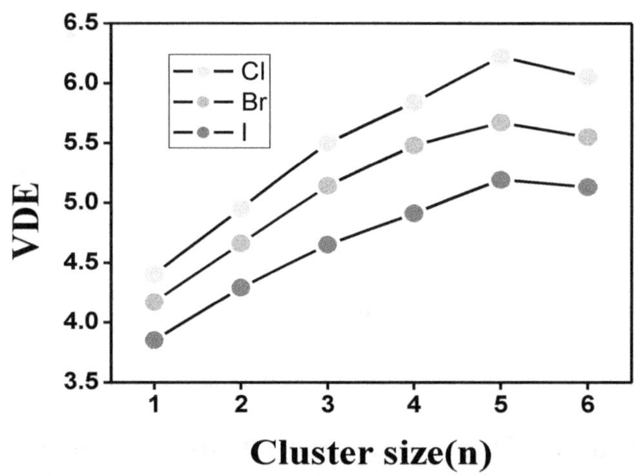

Figure 17. VDE versus cluster size for $(H_2O)_nX^-$

Table 12. VDE (eV) calculation of $(H_2O)_nX^-$ clusters

Cluster size	$(H_2O)_nCl^-$	$(H_2O)_nBr^-$	$(H_2O)_nI^-$
1	4.40	4.17	3.85
2	4.95	4.66	4.29
3	5.50	5.14	4.65
4	5.84	5.48	4.91
5	6.22	5.67	5.19
6	6.05	5.55	5.13

This is because of the strongest hydrogen bond being formed for the Cl^- and the weakest for the I^-. For a particular type of cluster, $(H_2O)_nX^-$ [X=Cl, Br, I], the VDE increased upto n=5 and then either saturates or further change becomes negligible. This is also explainable since as we increase the size of the cluster, the number of hydrogen bonds being formed with the central halide ion also increases, adding to the stability of the excess negative charge .However for n=6, the number of hydrogen bonds being formed with the central halide ion no more increases. The additional hydrogen bonds' being formed are within the water network and thereby does not affect further changes in VDE. The VDE data of the water-halide clusters are reported in Table 12.

D. NPA of $(H_2O)_nX^-$ clusters

In Figure 18, we report the amount of negative charge on the halide ion for each of the clusters considered. The general trend observed throughout the variation is that for a given size, magnitude of negative charge is the most for $(H_2O)_nI^-$ and least for $(H_2O)_nCl^-$ with $(H_2O)_nBr^-$ showing an intermediate value. This can be explained on the basis that the hydrogen bond strength is the most $(H_2O)_nI^-$ and least for $(H_2O)_nCl^-$ with $(H_2O)_nBr^-$ showing an intermediate value.

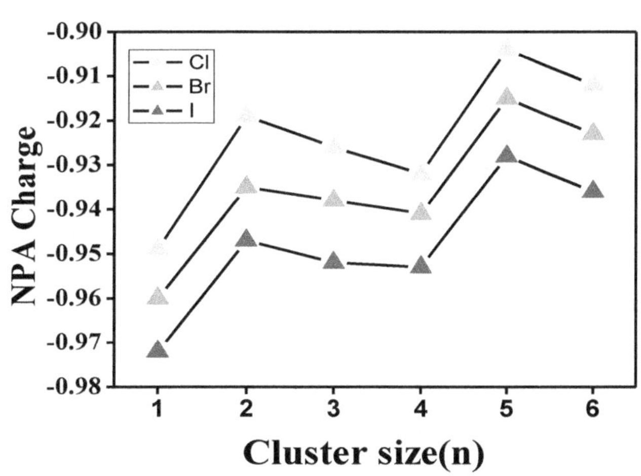

Figure 18. NPA Charge versus cluster size for $(H_2O)_nX^-$

Table 13. NPA charge calculation of $(H_2O)_nX^-$

Cluster size	$(H_2O)_nCl^-$	$(H_2O)_nBr^-$	$(H_2O)_nI^-$
1	-0.949	-0.960	-0.972
2	-0.919	-0.935	-0.947
3	-0.926	-0.938	-0.952
4	-0.932	-0.941	-0.953
5	-0.904	-0.915	-0.928
6	-0.912	-0.923	-0.936

This can be explained on the basis that the hydrogen bond strength is the most for $(H_2O)_nCl^-$ and least for $(H_2O)_nI^-$. With the increase in the strength of hydrogen bond greater transfer of negative charge from the Cl^- ion to the neighbouring atoms is expected. We also observe that for a particular type of cluster $(H_2O)_nX^-$ [X=Cl, Br, I], with the increase in size the magnitude of negative charge first decreases at n=2, then either decreases slightly or remains the same for n=3 and 4, then again rise for n=5 and then slightly dips for n=6. This feature is again due to the total strength of the hydrogen bonded interactions. Whenever the strength increases, the negative charge on the central halide ion decreases and vice-versa. The NPA data of water halide clusters are shown in Table 13.

3.4 Conclusions

We have shown that a strategy based on combined GA and DFT can be a viable tool to study structure and spectroscopy of water-halide clusters. This simple strategy of studying structure and spectroscopy supported by an empirical potential can answer many questions related to spectral shift associated with ion and size change of clusters. Various structures and spectra have been discussed for each of the hydrated halide ion clusters. The water-water interactions are compatible to the halide-water interactions; the water-water interaction is slightly stronger than the iodide-water interaction, but similar to bromide-water interaction, and slightly weaker than the chloride-water interaction. However as the number of water molecules increases, the water-water interactions are more important, and the surface structure is retained in X=Cl, Br and I. In chloride-water, bromide-water and iodide-water clusters surface-bound structures are more favoured. The O—H vibrational frequencies have been analyzed in terms of O—H_x, O—H_w and O—H_f in terms of proton donor or acceptor types of each water molecule. We have shown that the strategy based on Genetic Algorithm and Quantum mechanical calculations is able to broadly find out the differences in structural form and spectral pattern as the central ion changes from chloride to iodide.

3.5 References

[1] M. Yamashita, J. B. Fenn, J. Phys. Chem. **1984**, 88, 4671.

[2] J. V. Seeley, R. A. Morris, A. A. Viggiano, J. Phys. Chem. A **1997**, 101, 4598.

[3] J. V. Seeley, R. A. Morris, A. A. Viggiano, J. Phys. Chem. **1996**, 100,15821.

[4] H. Wincel, E. Mereand, A. W. Castleman, Jr., J. Phys. Chem. **1996**, 100,7488.

[5] U. Achatz, S. Joos, C. Berg, G. Neidner-Schatterburg, V. E. Bondybey, Chem. Phys. Lett. **1998**, 291, 459.

[6] C. J. Weinheimer, J. M. Lisy, J. Chem. Phys. **1996**, 105, 2938.

[7] O. M. Cabarcos, C. J. Weinheimer, J. M. Lisy, J. Chem. Phys. **1999**, 110,8429.

[8] G. Markovich, S. Pollack, R. Giniger, O. J. Cheshnovsky, Chem. Phys. **1994**,101,9344.

[9] C. G. Bailey, J. Kim, M. A. Johnson, J. Phys. Chem. **1996**, 100, 16782.

[10] M. S. Johnson, K. T. Kuwata, C. K. Wong, M. Okumura, Chem. Phys. Lett. **1996**,260, 551.

[11] M. A. Johnson, W. C. Lineberger, Techniques for the Study of Gas-Phase Ion Molecule Reactions; Wiley: New York, 1988.

[12] J. -H. Choi, K. T. Kuwata, Y. B. Cao, M. J. Okumura, Phys. Chem. A. **1998,** 102, 503.

[13] M. A. Johnson, M. L. Alexander, W. C. Lineberger, Chem. Phys. Lett. **1984**, 112,285.

[14] O. M. Cabarcos, C. J. Weinheimer, J. M. Lisy, S. S. Xantheas, J. Chem. Phys **1999**, 110, 5.

[15] J. R. Roscioli, E. G. Dicken, M. A. Johnson, J. Phys. Chem. A **2006**, 110, 4943.

[16] P. Ayotte, S. B. Nielsen, G. H. Weddle, M. A. Johnson, J. Phys. Chem. A. **1999**, 103, 10665.

[17] H. Kistenmacher, H. Popkie, E. Clementi, J. Chem. Phys. **1973**, 58, 1689.

[18] P. Lybrand, P. A. Kollman, J. Chem. Phys. **1985**, 83, 2923.

[19] P. Ayotte, G. H. Weddle, J. Kim, M. A. Johnson, J. Am. Chem. Soc.**1998,** 120, 12361.

[20] P. Ayotte, G. H. Weddle, J. Kim, M. A. Johnson, Chem. Phys. **1998**, 239, 485.

[21] P. Ayotte, J. A. Kelly, S. B. Nielsen, M. A. Johnson, Chem. Phys. Lett. **2000,** 316, 455.

[22] P. Ayotte, G. H. Weddle, M. A. Johnson, J. Chem. Phys. **1999,** 110, 7129.

[23] P. Ayotte, C. G. Bailey, M. A. Johnson, J. Phys. Chem. A. **1998,** 102, 3067.

[24] P. Ayotte, G. H. Weddle, J. Kim, M. A. Johnson, J. Phys. Chem. A. **1998**, 103, 443.

[25] J. A. Kelley, J. M. Weber, K. M. Lisle, W. H. Robertson, P. Ayotte, M. A. Johnson, Chem. Phys. Lett **2000**, 327, 1.

[26] S. A. Corcelli, J. A. Kelley, J. C. Tully, M. A. Johnson, J. Phys. Chem. A. **2002,** 106, 4872.

[27] M. Okumura, L. I. Yeh, J. D. Myers, Y. T. Lee, J. Chem. Phys. **1986,** 85, 2328.

[28] M. Okumura, L. I. Yeh, J. D. Myers, Y. T. Lee, J. Chem. Phys. **1990**, 94, 3416.

[29] H. E. Dorsett, R. O. Watts, S. S. Xantheas, J. Phys. Chem. A. **1999**, 103, 3351

[30] C. G. Bailey, D. J. Lavrich, D. Serxner, M. A. Johnson, J. Chem. Phys.**1996**, 105, 1807.

[31] C. G. Bailey, J. Kim, C. E. H. Dessent, M. A. Johnson, Chem. Phys. Lett. **1997**, 269, 122–127

[32] R. C. Dunbar, T. B. McMahon, D. Tholmann, D. S. Tonner, D. R. Salahub, D. Wei, J. Am. Chem. Soc. **1995**, 117, 12819.

[33] P. Kebarle, M. Arshadi, J. Scarborough, J. Chem. Phys. **1968**, 49, 817.

[34] M. Arshadi, R. Yamadgni, P. Kebarle, J. Phys. Chem. **1970**, 74, 1475.

[35] K. Hiraoka, S. Mizuse, S. Yamabe, J. Phys. Chem. **1988**, 92, 3943.

[36] A. N. Castleman, Jr., M. G. Brown, J. Phys. Chem. **1996**, 100, 12911.

[37] N. Takahashi, K. Sakaki, H. Tanida, I. Watanabe, Chem. Phys. Lett. **1995**, 246, 183.

[38] D. Serxner, C. E. H. Dessent, M. A. Johnson, J. Chem. Phys. **1996**, 105, 7231.

[39] G. Markovich, Giniger, R; M. Levin, O. Cheshnovsky, J. Chem. Phys. **1991**, 95, 9416.

[40] J. A. Kloepfer, V. H. Vilchiz, V. A. Lenchenkov, S. E. Bradforth, Chem. Phys. Lett. **1998**, 298, 120.

[41] J. A. Kloepfer, V. H. Vilchiz, V. A. Lenchenkov, A. C. Germaine, S. E. Bradforth, J. Chem. Phys. **2000**, 113, 6288.

[42] P. Wies, P. R. Kemper, M. T. Bowers, S. S. Xantheas, J. Am. Chem. Soc. **1999**, 121, 3531.

[43] J. K. Agreiter, A. M. Knight, M. A. Duncan, Chem. Phys. Lett. **1999,** 313, 162.

[44] M. A. Duncan, Annu. Rev. Phys. Chem. **1997**, 48, 69.

[45] D. E. Lessen, R. L. Asher, P. J. Brucat, J. Chem. Phys. **1990**, 93, 6102.

[46] D. C. Sperry, A. J. Midey, J. L. Lee, J. Quian, J. M. Farrar, J. Chem. Phys. **1999**, 111, 8469.

[47] M. Sanekata, F. Masaizu, K. Fuke, J. Chem. Phys. **1996**, 104, 9768.

[48] T. D. Vanden, B. Forinash, J. M. Lisy, J. Chem. Phys. **2002**, 117, 4628.

[49] J. Baik, J. Kim, D. Majumdar, S. K. Kim, J. Chem. Phys. **1999**, 110, 9116.

[50] S. S. Xantheas, Chem. Phys. **2000**, 258, 225.

[51] A. W. Castleman, Jr., K. H. Bowen, J. Phys. Chem. **1996**, 100, 12911.

[52] J. Z. Jortner, Phys. D At. Mol. Clusters **1992**, 24, 247.

[53] R. W. Gora, S. Roszak, Leszczynski, J. Chem. Phys. Lett. **2000**, 325, 7.

[54] S. S. Xantheas, J. Phys. Chem. **1996**, 100, 9703.

[55] J. E. Combariza, N. R. Kestner, J. Phys. Chem. **1994**, 98, 3513.

[56] D. Majumdar, J. Kim, K. S. Kim, J. Chem. Phys. **2000**, 112, 101.

[57] J. Kim, H. M. Lee, S. B. Suh, D. Majumdar, K. S. Kim, J. Chem. Phys. **2000**, 113, 5259.

[58] H. M. Lee, D. Kim, K. S. Kim, J. Chem. Phys. **2002**, 116, 5509.

[59] H. M. Lee, K. S. Kim, J. Chem. Phys. **2001**, 114, 4461.

[60] F. Yates, H. F. SchaeferIII, T. J. Lee, J. E. Rice, J. Am. Chem. Soc. **1988**, 110, 6327.

[61] W. L. Jorgensen, D. L. Severance, J. Chem. Phys. **1993**, 99, 4233.

[62] J. E. Combariza, N. R. Kestner, J. Jortner, Chem. Phys. Lett. **1993**, 203,423.

[63] J. E. Combariza, N. R. Kestner, J. Jortner, J. Chem. Phys. **1994**, 100,2851.

[64] S. S. Xantheas, T. H. Dunning, Jr., J. Phys. Chem. **1994**, 98, 13489.

[65] L. Perera, M. L. Berkowitz, J. Chem. Phys. **1994**, 100, 3085.

[66] L. Perera, M. L. Berkowitz, J. Chem. Phys. **1991**, 95, 1954.

[67] L. Perera, M. L. Berkowitz, J. Chem. Phys. **1993**, 99, 4236.

[68] L. Perera, M. L. Berkowitz, J. Chem. Phys. **1993**, 99, 4222.

[69] L. S. Sremaniak, L. Perera, M. L. Berkowitz, Chem. Phys. Lett. **1994**, 218, 377.

[70] L. Perera, M. L. Berkowitz, Z. Phys. D. **1993**, 26, 166.

[71] I. C. Yeh, L. Perera, M. L. Berkowitz, Chem. Phys. Lett. **1997**, 264, 31.

[72] L. X. Dang, B. C. Garrett, J. Chem. Phys. **1993**, 99, 2972.

[73] H. Gai, G. K. Schenter, L. X. Dang, B. C. Garrett, J. Chem. Phys. **1996**, 105, 8835.

[74] S. S. Xantheas, L. X. Dang, J. Phys. Chem. **1996**, 100, 3989.

[75] S. S. Xantheas, J. Am. Chem. Soc. **1995**, 117, 10373.

[76] S. Kim, H. F. Schaefer, III, J. Chem. Phys. **2007**, 126, 064301.

[77] D. K. Maity, J. Phys. Chem. A. **2002**, 106, 5716.

[78] A. K. Pathak, T. Mukherjee, D. K. Maity, Chem. Phys. Chem. **2010**, 11, 220.

[79] N. L. Jorgensen, J. Chandrasekhar, J. D. Madura, R. W.Impey, M. L. Klein J. Chem. Phys **1983**, 79, 926

[80] W. L. Jorgensen, J. Am. Chem. Soc. **1981**, 103, 335.

[81] D. J. Wales, M. P. Hodges, Chem. Phys. Lett. **1998**, 286, 65

[82] T. N. Truong, E. V. Stefanovich, Chem. Phys. **1997**, 218, 31.

[83] T. H. Dunning, Jr., J. Chem. Phys. **1989**, 90, 1007.

[84] R. A. Kendall, T. H. Dunning, Jr., R. J. Harrison, J. Chem. Phys. **1992**, 96, 6796.

[85] J. E. Combariza, N. R. Kestner, J. Phys. Chem. **1994**, 98, 3513.

[86] M. Masamura, J. Phys. Chem. A. **2002**, 106, 8925.

[87] M. Masamura, J. Chem. Phys. **2003**, 118, 6336.

[88] J. O. Jung, R. B. Gerber, J. Chem. Phys. **1996**, 105, 10332.

[89] M. Roeselova, G. Jacoby, U. Kaldor, P. Jungwirth, Chem. Phys. Lett. **1998**, 293, 309.

[90] H. E. Dorsett, R. O. Watts, S. S. Xantheas, J. Phys. Chem. A. **1999**, 103, 3351.

[91] L. X. Dang, D. E. Smith, J. Chem. Phys. **1993**, 99, 6950.

[92] S. J. Stuart, B. J. Berne, J. Phys. Chem. A. **1999**, 103, 10300.

[93] S. J. Stuart, B. J. Berne, J. Phys. Chem. **1996**, 100, 11934.

[94] R. Ayala, J. M. Martinez, R. R. Pappalardo, E. S. Macros, J. Phys. Chem. A**. 2000**, 104, 2799.

[95] J. Kim, K. S. Kim, J. Chem. Phys. **1998**, 109, 5886.

[96] W. H. Thompson, J. T. Hynes, J. Am. Chem. Soc. **2000**, 122, 6278.

[97] H. -Y. Chen, W. -S. Shen, J. Am. Chem. Soc. **2000**, 122, 7534.

[98] J. Cald-well, L. X. Dang, P. A. Kollman, J. Am. Chem. Soc. **1990**, 112, 9144.

[99] B. J. Mhin, J. Kim, K. S. Kim, Chem. Phys. Lett. **1993**, 216, 305.

[100] H. -Y. Chen, W. -S. Shen, Chem. Phys. Lett. **2001**, 335, 475.

[101] B. J. Mhin, S. Lee, J. Cho, K. Lee, K. S. Kim, Chem. Phys. Lett. **1992**, 197, 77.

[102] J. Kim, S. Lee, S. J. Cho, B. J. Mhin, K. S. Kim, J. Chem. Phys. **1995**, 102, 839.

[103] S. Lee, J. Kim, J. K. Park, K. S. Kim, J. Phys. Chem. **1996**, 100, 14329.

[104] H. M. Lee, J. Kim, S. Lee, B. H. Mhin, K. S. Kim, J. Chem. Phys. **1999**, 111, 3995.

[105] T. P. Lybrand, P. A. Kollman, J. Chem. Phys. **1985**, 83, 2923.

[106] S. -S. Sung, P. C. Jordan, J. Chem. Phys. **1986**, 85, 4045.

[107] S. Lin, P. C. Jordan, J. Chem. Phys. **1988**, 89, 7492.

[108] L. X. Dang, J. Chem. Phys. **1992**, 97, 2659.

[109] L. X. Dang, J. E. Rice, J. Cladwell, P. A. Kollman, J. Am. Chem. Soc. **1991**, 113, 2481.

[110] H. Gai, L. X. Dang, G. K. Schenter, B. C. Garrett, J. Phys. Chem. **1995**, 99, 13303.

[111] Y. Okuno, J. Chem. Phys. **1996,** 105, 5817.

[112] R. C. Dunbar, T. B. McMahon, D. Tholmann, D. S. Tonner, D. R. Salahub, D. Wei, J. Am. Chem. Soc. **1995**, 117, 12819.

[113] X. G. Zhao, A. L. Gonzalez, D. G. Truhlar, R. Steckler, J. Chem. Phys. **1991**, 94, 5544.

[114] C. -G. Iwata, S. Iwata, Chem. Phys. Lett. **1995**, 232, 72.

[115] S. J. Vaughn, E. V. Akhmatskaya, M. A. Vincent, A. J. Masters, I. H. Hiller, J. Chem. Phys. **1999**, 110, 4338.

[116] S. S. Xantheas, J. Am. Chem. Soc. **1995**, 117, 10373.

[117] N. L. Jorgensen, J. Chandrasekhar, J. D. Madura, R. W. Impey, M. L. Klein, J. Chem. Phys. **1983**, 79, 926.

[118] W. L. Jorgensen, J. Am. Chem. Soc. **1981**, 103, 335.

[119] D. J. Wales, M. P. Hodges, Chem. Phys. Lett. **1998**, 286, 65.

[120] P. Chaudhury, R. Saha, S. P. Bhattacharya, Chem. Phys. **2001**, 270, 277

[121] A. K. Pathak, T. Mukherjee, D. K. Maity, Chem. Phys. Chem. **2010**, 11, 220.

[122] J. Frisch et al. Gaussian 09 Revision B.01 (Pittsburgh: Gaussian, Inc.) **(2010)**

[123] D. E. Goldberg, Genetic Algorithm in Search, Optimization and Machine learning; Addison Wesley: Reading, MA, **1989.**

[124] J. Holland, Adaptation in Natural and Artificial Systems; MIT Press: Cambridge, MA, **1992**.

[125] M. Mitchell, An introduction to Genetic Algorithm, MIT Press: Cambridge, MA, **1998.**

[126] S. G. Neogi, P. Chaudhury, J. Comput. Chem. **2012**, 33, 629

Chapter 4

Structural and Spectroscopic studies of carbon dioxide clusters: a combined genetic algorithm and DFT based study

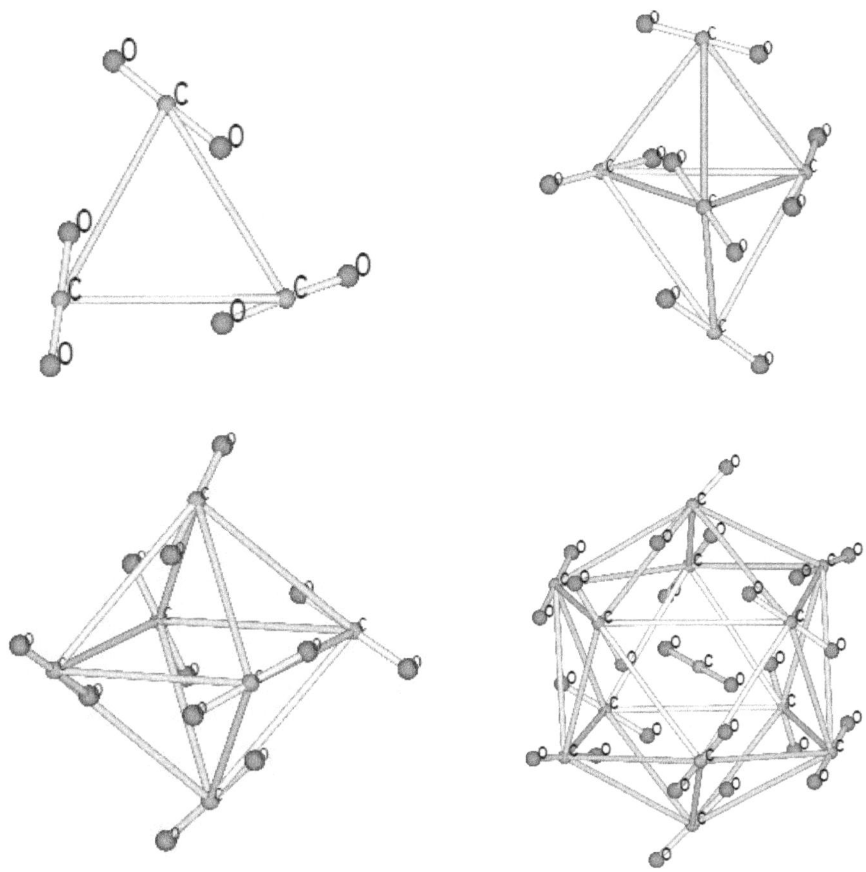

S. G. Neogi, S. Talukder, P. Chaudhury Structural Chemistry 2014, 25, 909

4.1 Introduction

The study of atomic and molecular clusters in which the chemical constituents are bound to each other by various types of forces ranging from weak van der Waals type of force to coulombic interactions have become an important area of study in contemporary research on experimental and theoretical chemical physics. These systems are interesting from the points of view of enormous structural diversity as well as the spectroscopic signatures that these subtle variations in structural forms contribute. Scientists look into these chemical systems from diverse angles.

CO_2 gas plays very important role in respiration and photosynthesis which are responsible for the existence of life on earth. Carbon dioxide has attracted considerable experimental and theoretical attention because under super-critical conditions CO_2 acts as a solvent in many chemical reactions and is widely used as dry cleaning agent. Solid carbon dioxide commonly known as dry ice is used as refrigerant. CO_2 finds its application in greenhouses as the main carbon source. Due to widespread applications, CO_2 clusters have been a subject of major experimental investigations [1–11]. Among the molecular clusters CO_2 rank second only to H_2O due to the importance of atmospheric CO_2 for the earth as well as the advantages of super-critical CO_2 as an industrial solvent[12,13].

There has been a persistence effort from both experimentalists and theoreticians to explain and understand the structural and spectroscopic properties of $(CO_2)_n$ clusters. High resolution infrared (IR) and Raman spectroscopic studies of Walsh et al. and Jucks et al.[3, 4] are worth mentioning. Experimentalists have used impressive techniques like electron beam diffraction [7], mass spectroscopy [8, 9] and direct absorption spectroscopy [10, 11] for larger CO_2 clusters. Using photoelectron spectroscopy, Harnes et al.[14] studied CO_2 clusters. High-resolution infrared (IR) spectroscopy of carbon dioxide have been studied by Oliaee et al. [15,16] and Dehghany et al.[17]. Theoreticians performed molecular dynamic studies [18–24, 25], ab-initio quantum chemical calculations [26–32] on $(CO_2)n$

clusters. Liu et al.[24] applied Monte-Carlo simulation technique in studying CO_2 clusters. Geometry Optimization of carbon dioxide clusters i.e.; $[(CO_2)n]$ for $4 \leq n \leq 40$ was performed by Takeuchi[25] using Lennard Jones (LJ) potential. Yeole et al.[33] have used Molecular Tailoring approach(MTA) to study structure, energetics and vibrational spectra of CO_2 clusters. Liu and Jordan [24] employed parallel-tempering Monte Carlo method to investigate finite temperature properties of $(CO_2)_n$ clusters. Very recently Temke et al.[34] studied thermodynamic properties of carbon dioxide clusters by M06-2X and Dispersion-Corrected B2PLYP-D theory.

Intermolecular potential functions for CO_2-CO_2 interaction are useful in predicting the structure of $(CO_2)n$ clusters. A theoretical study of possible structures of these systems can present interesting challenges to optimization techniques and procedures. A conventional ab-initio procedure is the most usual way followed by many research groups. A parallel strategy would be to define an empirical potential energy surface in terms of the inter-atomic distances present in the cluster system. These clusters, especially with the increase in size, can have multiple structures differing in energy and finding the lowest possible one or the global minimum of these systems face another stiff challenge to the theoreticians. This is the challenge of getting deeper minima. These potential energy surfaces are extremely rugged and the search for good structures can be a problem for any deterministic optimization methods. Hence there has been an increasing tendency to use truly global and stochastic optimization. Genetic Algorithm (GA) [35–39] is one such stochastic optimization technique. Ever since its introduction by Holland [40] it has found increasing use across disciplines ranging from engineering, physics and chemistry. GA [41] draws its working philosophy from the genetic process of selection, crossover and mutation. GA has been used to solve a host of problems ranging from structure elucidation of atomic and molecular clusters, solution of schrodinger equation, predicting geometries of polymeric materials etc. A new paradigm in studying chemical systems is also becoming popular, in which one uses GA in conjunction with standard quantum chemical procedures, especially Density Functional Theory (DFT) to study large cluster systems. The use of GA in elucidating structures of clusters has a rich and long history. Cluster systems ranging from Lennard Jones (LJ) clusters, Morse clusters, metallic clusters modelled by Gupta Potential and works of several research groups needs mention. The works of Deaven et al. [42,43], Pullan et al. [44–46], Niesse and Mayne [47–49], Johnston et al.[50,51], Michaelian et al.[52–55], Hartke et al.[56–59] and Ganguly Neogi et al.[60, 61] deserve special attention. In the present chapter we wish to test a GA+DFT method for studying $(CO_2)n$ clusters with n=2-13 and once the structures are obtained, we intend to get the infrared modes which is being calculated using DFT with ωB97XD[62–64] and M06-2X [65,66] being the functional used and 6-311++G(d,p) being the basis set used. Though M06-2X functional has been reported quite successful in incorporating dispersion interaction at least upto mid range, the dispersion corrected B2PLYP-D functional is adept at handling van der Waals bonded cluster systems. So we have also gone on to specifically calculate interaction energies with BSSE corrections for these cluster systems and reported in tabular form in results and discussion. All DFT calculations have been performed using Gaussian 09 package [67].

This two-pronged strategy of using an algorithm to build good quality initial guess structures followed by subsequent quantum chemical calculations used by us is by no means unique, and equally impressive report exists in literature using a molecular tailoring approach(MTA) [68–70, 71, 72–74] in combination with quantum chemical methods. It is noteworthy to recognise the contributions of Gadre and coworkers to the study on water [75], benzene [76], acetylene [77,78] and very recently N_2O cluster [79]. A special mention must be made on the applicability of MTA [75] as well as Monte Carlo Temperature Basin Paving approach [80] and MP2 methods to study large water clusters.

4.2 The Methodology

A. The Potential for the system

To model the interactions in $(CO_2)n$ clusters we have used the interaction potential made up of a combination of Lennard-Jones and Coulombic terms as given in the paper of Liu and Jordan[24]. The (CO_2) molecule is modelled such that five point charges are present on each monomer, the locations and values of which are given in Table 4.1. The origin is the carbon atom. The C-O bond length is 1.16Å. The $(CO_2)n$ cluster potential energy function is given by

$$U_{total} = \sum_i \sum_j \left(4\varepsilon \left[\left(\frac{\sigma_{ij}}{r_{ij}} \right)^{12} - \left(\frac{\sigma_{ij}}{r_{ij}} \right)^{6} \right] + \frac{q_i q_j}{r_{ij}} \right)$$

where the index i denotes the carbon and the sum run over all atoms(j) of the cluster system. The Lennard-Jones terms for CO_2-CO_2 interaction are listed in Table 4.2. The rigid five site model of $(CO_2)n$ cluster, though simple, incorporates all the major inter-atomic interactions present in the cluster system.

Table 4.1 Location of point charges in potential energy function for $(CO_2)n$

Site	Z(Å)	Q(e)
1	-1.5232	0.1216
2	-1.0663	-0.6418
3	0.0000	1.0404
4	1.0663	-0.6418
5	1.5232	0.1216

Table 4.2 LJ parameters in potential energy function for $(CO_2)n$

Atom pair	ε(K)	σ(Å)
C-C	26.3	2.824
C-O	75.2	3.026
O-O	44.5	2.925

B.Genetic Algorithm

The GA used by us has a population size of 20. The crossover used was a two point one, with the probability of crossover being 50% that is 50% of the strings were picked up randomly in a given generation for crossover. Mutation is obviously an infrequent event and the mutation probability was kept at 1% or roughly 1% of the individual variables in the solution pool were picked up to mutate. In GA run we have used 10000 generations. Since we used GA with a population size of 20, we obtained 20 solutions corresponding to each string used. These solutions were hierarchically arranged and the five lowest energy solutions from the set were each subjected to subsequent DFT calculations using the different functional as has already being stated.

C. Quantum Chemical Calculations

All the structures obtained from GA serves as input for the quantum chemical DFT calculations. We have used 6-311++G(d,p) basis set. All the $(CO_2)n$ clusters were explored at dispersion corrected ωB97XD[62–64] and hybrid meta exchange-correlation M06-2X[65,66] functional to locate minimum energy structures on the respective potential energy surfaces. Absence of imaginary frequencies in vibrational frequency calculation confirms that the structures are minima on the potential energy hyper surface. The cluster structures obtained from GA serves as input for DFT calculation. Infrared peaks of all the clusters are calculated at two DFT level of theory since spectroscopy is the signature of correct structure. All quantum chemical calculations were performed using GAUSSIAN 09 package [67]. The interaction energy for a particular $(CO_2)n$ cluster is calculated by using

$$\Delta E = E_n - n \times E_{mono}$$

where E_{mono} is the energy of monomer and E_n is the energy of a $(CO_2)n$ cluster. In addition to the two functionals mentioned, we have also employed B2PLYP-D functional to evaluate the interaction energies of the studied systems.

4.3. Results and Discussions

A. Structures of $(CO_2)n$ clusters

Our GA based search has yielded reasonable structures for $(CO_2)n$ clusters. All structures have been generated from randomly selected initial set of trial solutions and then subjecting to the GA operations till a maximum fitness solution is achieved. Once the GA evaluated structures were obtained DFT calculation [with basis set 6-311++G(d,p)] is performed to get the frequencies of the infrared (IR) modes. The structures returned by DFT calculation closely resembled those obtained by GA. We give an account of the structural features of the optimized clusters, after the GA stage and subsequent DFT calculation. The structures of $(CO_2)n$ clusters after GA+DFT calculation are shown in Figure 4.1. CO_2 dimer shows two isomers, the global minimum [Fig. 4.1(a)] corresponds to a planar slipped-parallel structure with centrosymmetric C_{2h} symmetry while the other isomer has a T-shaped geometry with C_{2V} symmetry [Fig. 4.1(b)]. Two isomers are possible for CO_2 trimer. A planar cyclic structure with C_{3h} symmetry [81] in which all the CO_2 monomers are equivalent and a twisted barrel shaped asymmetric top with C_2 symmetry. The cyclic C_{3h} structure is energetically lower. In the timer structure [Fig 4.1(c) and Fig 4.1(d)] the oxygen atoms are disposed in space in a manner so that the distance between them can be made as large as possible so as to minimize the destabilizing interaction between oxygen atoms. For tetramer both GA and DFT structures are essentially similar with minor differences in angular dispositions of oxygen atoms. A tetrahedral structure is obtained for $(CO_2)_4$ shown in Fig. 4.1(e). The non-coordinating oxygen atoms are arranged in such a manner with respect to each other that the destabilizing interaction between similar charged centres is minimized. The pentamer is shown in Fig 4.1(f). A trigonal bipyramid geometry possessing three CO_2 monomer in a triangular plane with two CO_2 occupying axial position.

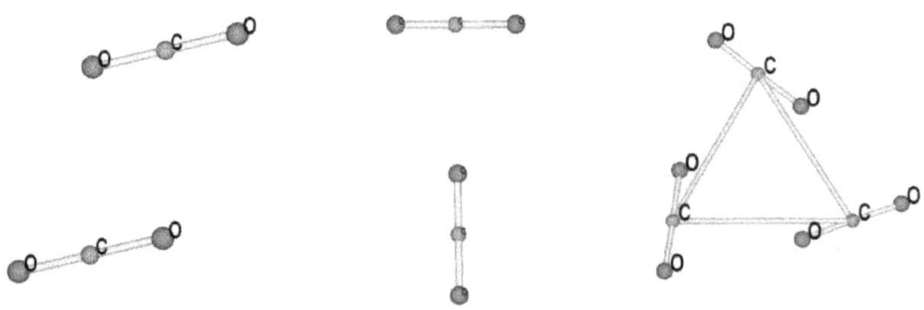

(a) n=2 (b) n=2 (c) n=3 (d) n=3

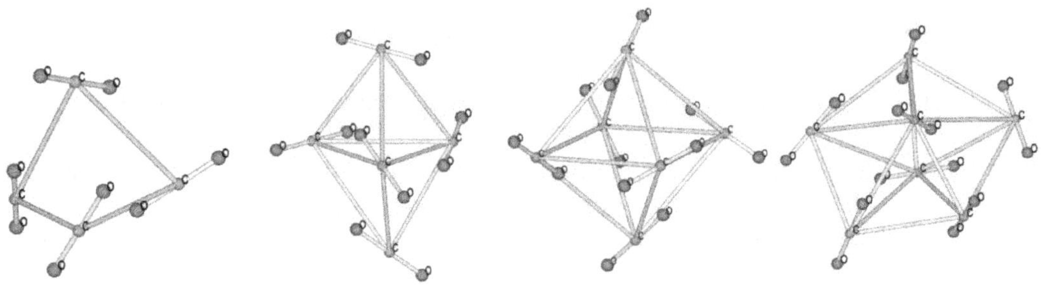

(e) n=4 (f) n=5 (g) n=6 (h) n=7

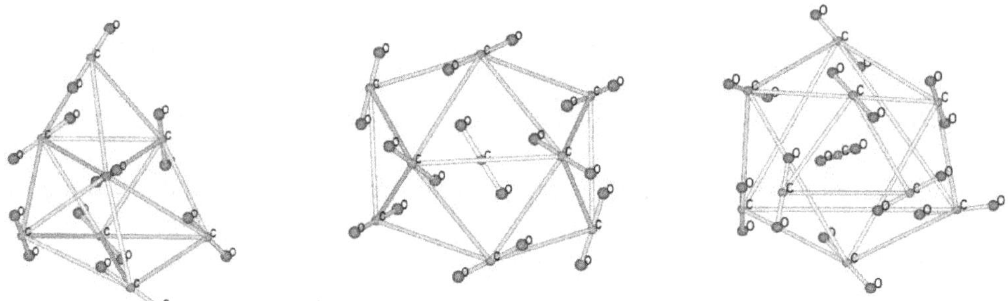

(i) n=8 (j) n=9 (k) n=10

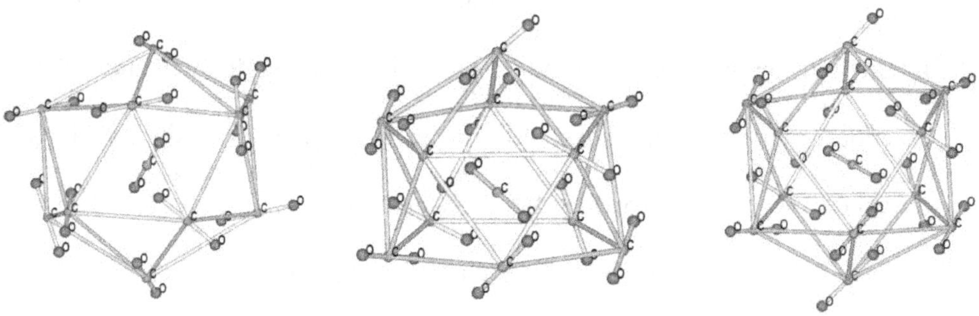

(l) n=11 (m) n=12 (n) n=13

Fig. 4.1 Structures of $(CO_2)n$ clusters after GA+DFT

Fig 4.1(g) show that the hexamer is a symmetric top (S_6 point group) formed by stacking two cyclic CO_2 trimers with co-incident symmetry axes but with top and bottom trimers rotated by 60 degree with respect to one another. For $(CO_2)_7$ heptamer [Fig 4.1(h)] the structure is unsymmetrical and consists of five monomers in a pentagonal ring with the remaining monomers above and below the ring. $(CO_2)_8$ and $(CO_2)_9$ both show unsymmetrical structure in both GA as well as DFT. Their structures are depicted in Fig 4.1 (i) and Fig.4.1 (j). The calculated structure of the decamer $(CO_2)_{10}$ appears to be an oblate symmetric top [Fig 4.1 (k)] whereas the carbon dioxide hen-decamer $(CO_2)_{11}$ shows an unsymmetrical structure [Fig 4.1(l)] as shown by (GA + DFT) calculation. As is the pattern with the other small size clusters, the non-coordinating oxygen atoms are arranged in such a manner with respect to each other that the destabilizing interaction between similar charged centres is minimized. The do-decamer $(CO_2)_{12}$ consists of one top monomer, two staggered pentagonal rings [Fig 4.1(m)]. The structure can be imagined of as a cup of eleven monomers with one monomer inside. A highly symmetric structure is obtained for the tridecamer $(CO_2)_{13}$ in both GA and DFT calculation. There is a single CO_2 monomer surrounded by a cage of 12 monomers whose carbon atoms lie at the vertices of a slightly distorted icosahedron, giving a symmetric top with S_6 point group symmetry [Fig 4.1(n)]. The interaction energy values of $(CO_2)n$ clusters obtained from DFT calculation are given in Table 4.3. BSSE corrected interaction energies are also calculated and given in Table 4.3. The interaction energy increases with the increase in cluster size. This is obvious as with the increase in cluster size the magnitude of van der Waals interaction increases rapidly and hence the difference in energy with the respective numbers of isolated monomer units increases.

Table 4.3 Interaction energy values (Kcal/mole) of $(CO_2)n$ from DFT [ωB97XD/6-311++G(d,p)] and [M06-2X/6-311++G(d,p)] and [B2PLYP-D/6-311++G(d,p)], w. BSSE refers to BSSE corrected, w.o. BSSE refers to without BSSE correction

CO_2 cluster	ωB97XD w.o BSSE	ωB97XD w.BSSE	M06-2X w.o.BSSE	M06-2X w.BSSE	B2PLYPD w.o.BSSE	B2PLYPD w.BSSE
2	-1.10	-0.93	-1.63	-1.43	-1.18	-1.08
3	-2.92	-2.41	-4.58	-4.24	-3.51	-3.13
4	-5.53	-4.42	-9.19	-7.83	-7.10	-6.02
5	-8.19	-6.51	-13.79	-11.64	-9.90	-8.77
6	-11.46	-9.02	-18.90	-15.73	-12.42	-10.71
7	-14.06	-10.96	-23.70	-19.84	-14.61	-12.50
8	-16.97	-13.20	-28.40	-23.67	-17.61	-14.30
9	-20.22	-15.67	-34.01	-28.17	-21.21	-16.10
10	-23.79	-18.26	-39.11	-32.33	-24.82	-17.91
11	-27.27	-20.93	-45.02	-37.00	-28.45	-19.72
12	-31.07	-23.70	-50.75	-41.48	-31.88	-21.53
13	-35.92	-27.16	-58.18	-47.31	-36.72	-23.35

B. Spectroscopy of $(CO_2)n$ clusters

The experimental C—O asymmetric stretch for the discrete CO_2 molecule is observed at 2350.8cm^{-1}.To obtain the infrared spectroscopic modes we have performed a normal coordinate analysis. The frequencies are harmonic and scaled by a factor of 0.95 for both M06-2X and ωB97XD. In the vibrational spectrum of CO_2 clusters, C—O asymmetric stretching vibration shows a blue shift with an increase in size from 2 to 13. The frequencies at the most intense peak correspond to the C—O asymmetric stretch for the most stable CO_2 clusters. These C—O asymmetric stretching frequencies generally reveal a blue shift with gradual increase of cluster size which is in agreement with the experimental results. The position of the most intense infrared peaks is depicted in Table 4.4.

Table 4.4 Vibrational frequencies of the strongest asymmetric C—O stretch for global minimum $(CO_2)n$ calculated at ωB97XD/6-311++G(d,p) and M06-2X/6-311++G(d,p) level of theory[16,81].

CO_2 cluster	Frequency(cm^{-1}) ωB97XD	Frequency(cm^{-1}) M06-2X	Experimental Ref. [15]
2 C_{2h}	2345.4	2354.2	2350.7
3 C_{3h}	2346.2	2346.8	2364.0
4	2341.5	2354.3	
5	2343.4	2355.9	
6	2341.3	2354.9	2353.6
7	2347.1	2359.5	2356.3
8	2346.5	2360.4	2357.6
9	2352.1	2358.3	2358.7
10	2350.9	2362.5	2361.2
11	2356.2	2363.7	2361.8
12	2358.6	2364.6	2364.3
13	2357.5	2365.0	2367.7

The frequencies are scaled by a factor of 0.95. There is a very close correspondence between the experimental frequencies and the frequencies obtained for the corresponding carbon-dioxide clusters. The correspondence of the experimental frequencies [15] to the calculated frequencies is nearly 99%. As seen in Table 4.4, the vibrational frequency for asymmetric C—O stretch gets slightly blue shifted as one moves from the smaller $(CO_2)n$ clusters with n=2,3 to the relatively larger clusters with n=12,13. This has already been reported by both experimentalists [15] and theoreticians [33] in their earlier works. What we find in our study of the structures of $(CO_2)n$ clusters is that as the size of the clusters increases, the two individual bonds in each of the CO_2 molecule get placed in two dissimilar chemical environments. As evident from the pictorial representation of the cluster structure[like Fig. 4.1(n) for $(CO_2)_{13}$], while one of the C-O bond gets trapped in the inner structural cavity and gets interacted with other CO_2 units due to van der Waals interactions, the other C—O bond is seen to be projecting out of the main cluster cavity environment. This bond which is projecting out experiences minimal van der Waals interaction and gets significantly shortened. On account of the shortening of the C—O bonds with gradual increase in size of the cluster a blue shift in the frequency is observed. A shorter bond would obviously lead to a vibration with a greater force constant and hence higher frequency.

C. Carbon-Carbon non-covalent interactions in $(CO_2)n$ clusters

Table4.5 Variation of C—C bond length of $(CO_2)n$ Calculated at M06-2X/6-311++G(d,p)

$(CO_2)n$ cluster	Average C—C bond length(Å)
2	3.41
3	3.44
4	3.57
5	3.60
6	3.64
7	3.53
8	3.58
9	3.61
10	3.73
11	3.69
12	3.74
13	3.64

It is expected that the choice of M06-2X functional for doing the final quantum chemical calculation at the DFT level after pre-optimization by GA should give C-C interaction distances with a much higher degree of accuracy. The inbuilt theoretical tool to incorporate medium range dispersion interactions must be the reason for the success of this functional and one expects the results to be comparable or in good agreement with higher level theoretical calculations like CCSD(T). The average C-C interaction lengths are reported in Table 4.5. The variations do not show a monotonic rise or fall. However one can notice an increase from the small size clusters (like the dimer and trimer) to the larger ones (dodecamer and tridecamer). In particular we would like to emphasize the good agreement between our calculated value for the dimer (3.41Å) and the value of 3.46Å reported in literature[34]. There has also been a report [34] of a value of 3.38Å for the dimer using a DFT calculation with the same functional. This is also in good agreement with the value obtained by us.

4.4. Conclusion

We have shown that the stochastic optimization technique Genetic Algorithm in conjunction with density functional theory can lead to correct determination of global minimum structure for $(CO_2)n$ clusters. The computed vibrational frequencies for an asymmetric C—O stretch generally exhibit a blue shift with increasing cluster size. This observation is in agreement with the experimental results. The correspondence of the computed frequencies to the experimental frequencies is nearly 99% at M06-2X/6-311++G(d,p) level of calculation. We have shown that using the GA structures as input for subsequent DFT calculation a quicker convergence can be achieved which reduces the computational cost. The use of hybrid meta GGA Minnesota functional M06-2X provides very accurate study for $(CO_2)n$ clusters where van der Waals dispersion interaction is important which is evident from the position of the infrared peaks. This strategy based on combined GA and DFT can be a viable tool to study structure and spectroscopy of $(CO_2)n$ clusters. In the infrared (IR) spectrum of CO_2 clusters, C—O asymmetric stretching vibration shows a blue shift with an increase in size from 2 to 13. This simple strategy of studying structure and spectroscopy supported by an empirical potential can answer many questions related to spectral shift associated with size change of clusters.

4.5 References

[1] A. S. Gopalan, C. M. Wai and H. K. Jacobs, ACS Symposium series, **2003**, p860

[2] W. Stafford, Nature **2007**, 448, 526

[3] M. A. Walsh, T. H. England, T. R. Dyke and B. J. Howard, Chem. Phys. Lett **1987**, 142, 265

[4] K. W. Jucks, Z. S. Huang, D. Dayton, R. E. Miller and W. J. Lafferty, J. Chem. Phys. **1987**, 86, 4341

[5] G. T. Fraser, A. S. Pine, W. J. Lafferty and R. E. Miller, J. Chem. Phys. **1987**, 87, 1502

[6] M. J. Wedia and D. J. Nesbit, J. Chem. Phys. **1996**, 105, 10210

[7] G. Torchet, M.-F. de Feraudy, A. Boutin and A. H. Fuchs, J. Chem. Phys. **1996**, 105, 3671

[8] Y. Negishi, T. Nagata and T. Tsukuda, Chem. Phys. Lett **2002**, 364, 127

[9] O. Ingolfsson and A. M. Wodtke, Chem. Phys. Lett **2002**, 360, 415

[10] A. Bonnamy, R. Georges, E. Hugo and R. Sugnorell, Phys. Chem. Chem. Phys. **2005**, 7, 963

[11] A. Bonnamy, R. Georges, A. Benidar, J. Boissoles, A. Canosa and B. R. Rowe, J. Chem. Phys. **2003**, 118, 3612

[12] J. M. Desimone, Science **2002**, 297, 799

[13] S. Balasubramanian, A. Kohlmeyer and M. L. Klein, J. Chem. Phys. **2009**, 131, 144506

[14] J. Harnes, M. Winkler, A. Lindblad, L. J. Saethre and K. J. Borve, J. Phys. Chem. A **2011**, 115, 10408

[15] J. N. Oliaee, M. Dehghany, A. R. W. Mckellar and N. Moazzen-Ahmadi, J. Chem. Phys **2011**, 135, 044315

[16] J. N. Oliaee, M. Dehghany, N. Moazzen-Ahmadi and A. R. W. Mckellar, Phys. Chem. Chem. Phys **2011**, 13, 1297

[17] M. Dehghany, A. R. W. Mckellar, M. Afshari, N. Moazzen-Ahmadi, Mol. Phys **2010**, 108, 2195

[18] C. S. Murthy, S. F. O'Shea and I. R. McDonald, Mol. Phys **1983**, 50, 531

[19] B. W. van der Waals, J. Chem. Phys **1983**, 79, 3948

[20] J. S. Muenter, J. Chem. Phys **1991**, 94, 2781

[21] G. Steinbrunner, A. J. Dyson, B. Kirchner and H. Huber, J. Chem. Phys **1998**, 109, 3153

[22] J.-B. Maillet, A. Boutton, S. Buttefey, F. Clavo and A. H. Fuchs, J. Chem. Phys **1998**, 109, 329

[23] J.-B. Maillet, A. Boutin and A. H. Fuchs, J. Chem. Phys **1999**, 111, 2095

[24] H. Liu and K. D. Jordan, J. Phys. Chem. A **2003**, 107, 5703

[25] H. Takeuchi, J. Phys. Chem. A **2008**, 112, 7492

[26] R. Bukowski, J. Sedlej, B. Jeziorski, P. Jankowski, K. Szalewicz, S. A. kucharski, H. L.Williams and B. M. Rice, J. Chem. Phys **1999**, 110, 3785

[27] S. Tsuzuki, W. klopper and H. P. Luthi, J. Chem. Phys **1999**, 111, 3846

[28] K. V. Jose and S. R. Gadre, Int. J. Quant. Chem **2009**, 109, 2238

[29] K. V. Jose and S. R. Gadre, J. Chem. Phys **2008**, 128, 124310

[30] V. Ganesh, R. K. Dongare, P. Balanarayan and S. R. Gadre, J. Chem. Phys 2006, 125, 104109

[31] A. P. Rahalkar, V. Ganesh and S. R. Gadre, J. Chem. Phys **2008**, 129, 234101

[32] J. D. McMahon, J. R. Lane, J. Chem. Phys **2011**, 135, 154309

[33] S. D. Yeole, N. Sahu and S. R. Gadre, Phys. Chem. Chem. Phys **2012**, 14, 7718

[34] K. H. Lemke, T. H. Seward, Chem. Phys. Lett **2013**, 573, 19

[35] D. M. Deaven, K. M. Ho, Phys. Rev. Lett. **1995**, 75, 288

[36] P. Chaudhury, R. Saha, S. P. Bhattacharya, Chem. Phys. **2001**, 270, 277

[37] P. Chaudhury, S. P. Bhattacharya, Chem. Phys **1999**, 241, 313

[38] C. Roberts, R. L. Johnston, N. T. Wilson, Theo. Chem. Acc. **2000**, 104, 123

[39] D. E. Goldberg, Genetic Algorithm in Search, Optimization and Machine learning; Addison Wesley: Reading MA, **1989**

[40] J. Holland, Adaptation in Natural and Artificial Systems, MIT Press, **1992**

[41] M. Mitchell, An introduction to Genetic Algorithm, MIT Press, Cambridge, MA, **1998**

[42] D. M. Deaven, J. R. Morris, K. M. Ho, N. Tit, Chem. Phys. Lett. **1996**, 256, 195

[43] D. M. Deaven, K. M. Ho, Phys. Rev. Lett. **1995**, 75, 288

[44] W. J. Pullan, J. Chem. Inf. Modelling. **1997**, 37, 1189

[45] W. J. Pullan, Comp. Phys. Comm. **1997**, 107, 137

[46] W. J. Pullan, J. Comput. Chem. **1997**, 18, 1096

[47] J. A. Niesse, H. R. Mayne, Chem. Phys. Lett. **1996**, 261, 576

[48] J. A. Niesse, H. R. Mayne, J. Chem. Phys. **1996**, 105, 4700

[49] J. A. Niesse, H. R. Mayne, J. Comput. Chem. **1997**, 18, 1233

[50] R. L. Johnston, Dalton. Transc. **2003**, 4193

[51] R. Ismail, R. L. Johnston, Phys. Chem. Chem. Phys. **2010**, 12, 8607

[52] K. Michaelian, Chem. Phys. Lett. **1998**, 293, 202

[53] K. Michaelian, Am. J. Phys. **1998**, 66, 231

[54] K. Michaelian, I. L. Garzon, M. R. Beltran, P. Ordejon, D. Portal-Sanchez, J. M. Soler, E. Artacho, Phys. Rev. Lett. **1998**, 81, 1600

[55] I. L. Garzon, N. Rendon, K. Michaelian, Phys. Rev. B. **1999**, 60, 2000

[56] B. Hartke, J. Phys. Chem. **1993**, 97, 9973

[57] B. Hartke, Chem. Phys. Lett. **1995**, 240, 560

[58] B. Hartke, Chem. Phys. Lett. **1996**, 258, 144

[59] B. Hartke, Phys. Chem. Chem. Phys. **2003**, 5, 275

[60] S. G. Neogi, P. Chaudhury, J. Comput. Chem. **2012**, 33, 629

[61] S. G. Neogi, P. Chaudhury, J. Comput.Chem. **2013**, 34, 471

[62] S. Grimme, J. Antony, S. Ehrlich and H. Krieg, J. Chem. Phys. **2010**, 132, 154104

[63] S. Grimme, J. Comput. Chem. **2006**, 27, 1787

[64] J. Chai and M. Head-Gordon, Phys. Chem. Chem. Phys., **2008**, 10, 6615

[65] Y. Zhao, D. G. Truhlar, Theor. Chem. Acc **2008**, 120, 215

[66] Y. Zhao, D. G. Truhlar, J. Chem. Theor. and Comput **2008**, 4, 1849

[67] Gaussian 09, Revision B.01, M. J. Frisch, Gaussian, Inc., Wallingford CT, **2010**.

[68] S. R. Gadre, K. V. JovanJose, A. P. Rahalkar, J. Chem. Sci. **2010**, 122, 47

[69] S. D. Yeole, S. R. Gadre, J. Chem. Phys. **2011**, 134, 08411

[70] S. D. Yeole, S. R. Gadre, J. Phys. Chem. A **2011**, 115, 12769

[71] N. Sahu, S. D. Yeole, S. R. Gadre J. Chem. Phys **2013**, 138, 104101

[72] S. R. Gadre, R. N. Shirsat, A. C. Limaye J. Phys. Chem. A **1994**, 98, 9165

[73] K. Babu, S. R. Gadre, J. Comput. Chem **2003**, 24, 484

[74] V. Ganesh, R. K. Dongare, P. Bandyopadhyay and S. R. Gadre, J. Chem. Phys. **2006**, 125, 104109

[75] J. P. Furtado, A. P. Rahalkar, S. Shanker, P. Bandyopadhyay and S. R. Gadre, J. Phys. Chem. Lett **2012**, 3, 2253

[76] A. S. Mahadevi, A. P. Rahalkar, S. R. Gadre, R. N. Shastry J. Chem. Phys **2010**, 133, 164308

[77] A. P. Rahalkar, B. K. Misra, V. Ramanathan, S. R. Gadre Theor. Chem. Acc **2011**, 130, 491

[78] A. P. Rahalkar, S. D. Yeole, S. R. Gadre Theor. Chem. Acc. **2012**, 131, 1095

[79] S. D. Yeole, N. Sahu and S. R. Gadre, J. Phys. Chem. A **2013**, 117, 8591

[80] S. Shanker, P. Bandyopadhyay, J. Phys. Chem. A **2005**, 115, 11866

[81] M. Dehghany, M. Afshari, N. Moazzen-Ahmadi, and A. R.W. McKellar, J. Chem. Phys. **2008**, 128, 064308

Chapter 5

Structural, Spectroscopic Aspects, and Electronic Properties of neutral lattice-like (MgO)n Clusters: A study Based on a blending of DFT with stochastic algorithms inspired by natural process

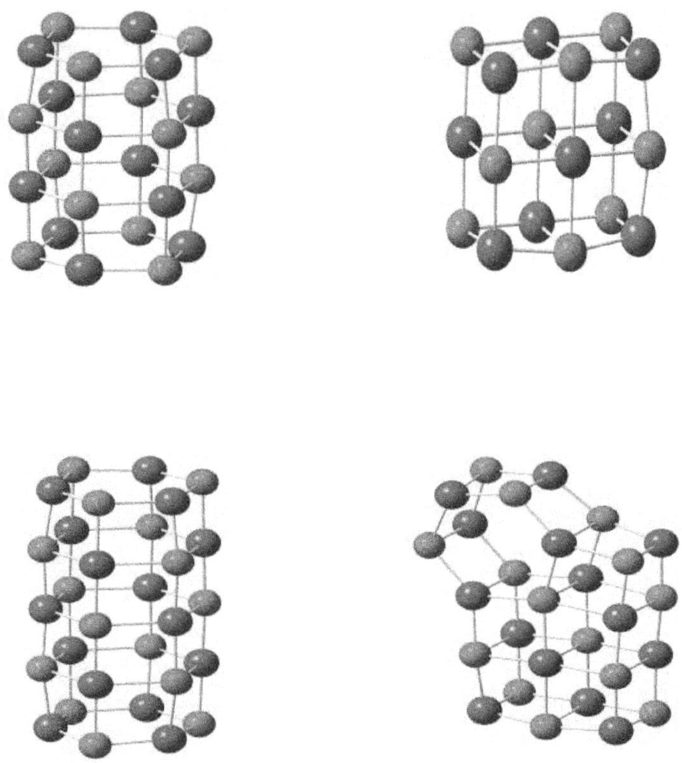

S. G. Neogi, P. Chaudhury Structural Chemistry 2014, 25, 1229

5.1 Introduction

Clusters are aggregates between a few and thousands of atoms and molecules. Cluster structure results in nanotubes, rods and cages from which the condensed matter properties arise. Nanoscale clusters [1–3] are an important topic in current scientific research because nanotubes, rods and cages are interesting structural motifs and are candidates for electrical and mechanical parts in nanodevices. Hence studies of cluster properties as a function of size has attracted researchers in recent years. The study of the properties of metal-oxide nanostructures has emerged as a research field of great interest in recent times. The evolution of these properties with cluster size is an important challenge to both experimentalists and theoreticians.

Magnesium oxide (MgO) is a simple metal oxide which is widely used in technological applications. Bulk MgO is an inert material with a high melting point and is also a wide bandgap (7.8eV) insulator. The widespread application of MgO in nanotechnology helps in understanding the transition from small clusters to bulk materials. MgO is a very important metal oxide supporter of catalysis. Recently, nanotubes, rods and cages of MgO [4, 5] nanostructures have been successfully explored and their physical properties have been investigated [6, 7]. MgO nanorods and nanotubes were fabricated by simple thermal methods [8, 9]. Multiple photon laser ionization mass spectrometry measurements [10] have revealed that tubular structures are formed by stacking of hexagonal $(MgO)_3$ units in addition to cubic motif. Saunders [11, 12] studied the stability of $(MgO)_3$ clusters by performing experiments in quadrupole mass spectrometry. Size selective infrared spectra of (MgO)n clusters with n=3-16 have been studied in gas phase with a tunable IR-UV two color ionization scheme by Haertelt et al.[13]. Moukouri and Nogurea[14] have performed semi-empirical tight binding calculations on (MgO)n clusters with n=1-6. Several theoreticians have carried out more accurate ab initio calculations on (MgO)n clusters. In this connection the studies of Recio and co-workers [15, 16], Jain et al.[17], Hong et al.[18], Malliavin and Coudray[19] and de la Puente et al.[20] are worth mentioning. Many theoreticians applied simple ion (pair) potentials [20–24] in conjunction with several optimization procedures [7, 25–27] to elucidate global structures of (MgO)n clusters. Quantum Chemical methods [27–29] are used by theoreticians to study size dependent structural and electronic properties of (MgO)n clusters. Experimentalists have used matrix isolation spectroscopy to study $(MgO)_2$ [30] cluster. For magnesium oxide clusters of larger sizes, IR resonance enhanced multiple photon ionization (IR-REMPI) spectra [31] is reported in literature. Very recently Kwapien et al.[32] employed messenger tagging technique to study IR spectra of small cationic $(MgO)^{+}_{2-7}$ clusters. Theoretical and experimental studies on the properties of neutral (MgO)n clusters have revealed some interesting phenomena in comparison to the bulk. Zhang et al.[27] employed molecular dynamics simulation to study (MgO)n clusters with n=18, 21 and 24. Dong et al.[7] explored the stability of $(MgO)_{3n}$ (n=2-10) clusters using Genetic Algorithm in association with Density Functional Theory(DFT). There is also an impressive collection of work on magnesium oxide system by Pacchioni and co-workers [33–36] and deserves special acknowledgement.

A theoretical study of possible structures of these systems can present interesting challenges to optimization techniques and procedures. Deciphering the correct structure would essentially mean finding the correct set of interatomic distances for which the potential energy is a minimum. These clusters, for a given specific size (n), because of being held together by weakly interacting forces can show up a large number of possible structures within a small energy range. Since the quantum chemical methods such as the density functional theory (DFT) and MP2 methods are not suitable for large size clusters due to the huge computational cost and numerous local minima [37–40], developing efficient global optimization algorithms to efficiently search out empirical potential energy surfaces is an important alternate way to understand the structures and properties of these clusters [41–43]. As the presence of multiple minima configurations are a strong possibility in these systems, the traditional gradient based search techniques employed commonly in quantum chemical calculations might not be able to find out the global minimum configuration with certainty. However, the so called non-gradient search techniques employing principles of stochasticity like simulated annealing[44,45], genetic algorithm[46–48], parallel tempering[49], basin hopping[50–57] algorithm and dynamical lattice searching (DLS) method[42-62] have been used effectively to search out the global minimum with certainty with the special power of being able to search out the global geometry independent of the starting guess which is being used to start the exploration of the potential energy surface. Here we use two techniques, namely Simulated Annealing and Genetic Algorithm in combination with density functional theory to elucidate global minimum structures of (MgO)n clusters with n=2-15.

We in this present chapter would like to use SA/GA as the preferred stochastic technique to search out an empirically defined potential energy surface(PES) and then use the SA/GA optimized structures as inputs for DFT study. Our contention is that since the empirical PES is one of high quality, the global minimum defined on it will be very close to the minimum found by DFT calculations. This we believe is a much smarter strategy than doing a quantum chemical calculation from the outset. As the initial structures for DFT calculations are very close to the minimum (predicted by DFT), convergence will be extremely quick as opposed to a strategy based on a pure DFT calculation. In the present chapter, for neutral (MgO)n clusters, we would like to see how the IR spectroscopic modes evolve with the change in cluster size. We would also like to address how the properties like cluster formation energy, second-difference energy, HOMO-LUMO gap, vertical ionization potential, Singlet-Triplet gap and non-linear optical properties vary with change in size of the clusters.

5.2. The Methodology

A. The Potential for the system

In order to describe the interactions that exist between all the atoms constituting a cluster of the type (MgO)n where n is the number of Mg atoms, the interaction between the ions is described by an empirical potential that includes long range electrostatic coulomb energy and a repulsive Born-Mayer potential:

$$V_{ij} = \frac{q_i q_j e^2}{4\Pi\varepsilon_o r_{ij}} + B_{ij}\, exp\left(\frac{-r_{ij}}{\rho_{ij}}\right) \qquad (1)$$

where, q_i and q_j are the charges on ions i and j respectively and r_{ij} is the distance between the ions 'i' and 'j'. B_{ij} and ρ_{ij} are Born-Mayer energy and distance scaling parameters respectively. The total potential of a cluster of N ions is then obtained by summing over all pairs of ions:

$$V_{clus} = \sum_{i=1}^{N-1}\sum_{j>i}^{N} V_{ij} \qquad (2)$$

The charges are Mg=1.00|e| and O=-1.00|e| and the values of B_{ij} and ρ_{ij} are given in Table 5.1. The parameters used in the potential energy function are taken from the paper of C. Roberts and R. L. Johnston [25]. Since the present work focuses on the identification of lattice like structures having a strong propensity for being cubic, the potential used has one unit of charge on the cation and anion, of course with different signs. The increase in the magnitude of charge on the ions will take the structures to a more spherical or ring like ones as has been reported earlier. Hence the justification of the magnitude of charge used is based on our emphasis on the study of lattice like structures as opposed to the spherical ones.

Table 5.1. Parameters used in potential energy function of (MgO)n Clusters

parameter	value
B_{ij}(Mg-O)	821.6 eV
B_{ij}(O-O)	22764 eV
ρ_{ij}(Mg-O)	0.3242 Å
ρ_{ij}(O-O)	0.1490 Å

Table 5.2: Parameters used in different optimization schemes

SA	Initial annealing temp	1.0
	Annealing schedule	1%
	Order of maximum change in a move	1%
GA	Crossover probability	100%
	Mutation probability	5%
	Mutation amplitude	1%

B. Computational Details

The problem we are studying in the present chapter is to get the right structure of the $(MgO)_n$ system, such that the potential energy of the system is a minimum. So essentially, the solution of the problem is to find optimum coordinates for all the atoms in the cluster. We have used Simulated Annealing and Genetic Algorithm in finding out the structures of (MgO)n cluster systems. The details of Simulated Annealing and Genetic Algorithm have been described in Chapter1 of the thesis. The SA and GA parameters are given in Table 5.2. The Cartesian coordinates obtained from SA/GA simulation are used as input in geometry optimization using DFT. All of the calculations reported in the chapter were obtained with GAUSSIAN 09[95] suite of programs. Electron correlation was partially taken into account using the hybrid functional usually denoted as B3LYP and the standard 6-311++G(d,p) basis set for Magnesium and Oxygen. The geometries of (MgO)n clusters were fully optimized with the hybrid B3LYP—DFT method. All the stationary points were characterized by harmonic vibrational frequency analysis as minima (Number of imaginary frequency=0). In order to explain the relative stabilities of these clusters, we have calculated the formation energy $E_f(=En/n - E_1)$ of the lowest lying structures. E_1 and E_n are the total energies of $(MgO)_1$ and (MgO)n clusters. TDDFT/B3LYP calculations are performed to evaluate vertical excitation energies to the lowest singlet states. The vertical ionization potentials (IP_v) are calculated by single point B3LYP/6-311++G(d,p) calculations at optimized neutral cluster geometries for singly positively charged clusters.

$$IPv= E[(MgO)_n]^+ - E[(MgO)_n]$$

Here $E[(MgO)n]^+$ and E[(MgO)n] are the total energies of charged and neutral clusters respectively without zero point correction. The Second difference in energy (Δ^2E) is given by $[(\Delta^2E = 2E_n-(E_{n+1}+E_{n-1})]$ where E_{n+1} and E_{n-1} are the total energies of $(MgO)_{n+1}$ and $(MgO)_{n-1}$ clusters.

5.3. Results and Discussion

A. Structures of (MgO)n clusters

The initial structure of (MgO)n clusters using the empirical potential were evaluated using both the stochastic techniques of SA and GA. Our objective of using both these techniques was to see whether both were equally capable of handling cluster systems and also as a crosscheck of the geometries obtained by either SA or GA. It happens that both these techniques were able to locate the global minimum geometry. Since both the techniques arrive at the same structures it signifies that the structures are indeed of high quality and are perfect candidates to be used as inputs in subsequent DFT calculations. The global and local (MgO)n clusters involving SA/GA followed by DFT are given in Fig. 5.1, 5.2 and 5.3. $(MgO)_2$ shows a square planar structure(D_{2h}) in which each Mg atom bonds with two oxygen atoms in an alternating way. The lowest energy structure of $(MgO)_3$ cluster is a planar hexagonal ring with each atom having a coordination number two. $(MgO)_4$ shows a distorted cubic structure as the global minimum where magnesium and oxygen atoms are placed alternately. The global minimum structure of $(MgO)_5$ is quite different. It is a structure with Cs point group showing both cubic as well as hexagonal ring motifs. The structure can be obtained by adding one MgO unit to the $(MgO)_4$ cluster. The structure contains a pair of two-coordinated O and Mg atoms while the other two atoms have 3-fold coordination. The global minimum (D_{3d}) structure of the hexamer $(MgO)_6$ is formed by stacking of two staggered hexagonal rings with Mg and O atoms located alternatively as shown in fig. 5.1. The lowest energy structure of the heptamer $(MgO)_7$ possess a cage like structure of C_{3V} point group. All the Mg and O atoms are 3 fold coordinated in this structure. These structures $(MgO)_2$ -- $(MgO)_7$ agrees with the lowest energy structure found by Hong et al.[18]. The lowest energy conformer of $(MgO)_8$ is characterized by a cage-like structure involving S_4 symmetry. The structure is formed by a $(MgO)_6$ cluster that is capped with a $(MgO)_2$ unit. The geometry of the global minimum of $(MgO)_9$ is characterized by three hexagonal $(MgO)_3$ motifs stacked each other alternately. The structure shows (D_{3h}) point group symmetry. The lowest energy structure of the decamer is shown in Fig. 5.2. The structure can be obtained by capping $(MgO)_2$ unit to the side of the $(MgO)_8$ cluster. The structural motif is of C_2 symmetry. The hendecamer $(MgO)_{11}$ cluster shows four hexagonal $(MgO)_3$ units fused together with two $(MgO)_2$ as given in Fig. 5.2. The geometry of the global minimum is characterized by a mixed structure formed by a stacking of hexagonal rings in conjunction with a cuboidal bulk like part. The lowest energy isomer of the dodecamer $(MgO)_{12}$ obtained is a hexagonal tube structure of D_{3d} symmetry formed by stacking of four $(MgO)_3$ hexagons. The structural motif of $(MgO)_{13}$ is depicted in Fig. 5.3. The structure consists of 3×2×2 lattice with an octagonal $(MgO)_4$ ring stacked above the O-centered 3×3 face. The lowest energy structure of $(MgO)_{14}$ obtained from our SA/GA+DFT study looks somewhat similar to a local structure of $(MgO)_{11}$. The motif is characterized by a mixed structure formed by stacking of two hexagonal rings in association with a cuboidal bulk like part. The structure belongs to (C_1) point group. The global minimum of $(MgO)_{15}$ is again characterized by hexagonal tube of D_{3h} symmetry where the magnesium and oxygen atoms are placed alternately. Most of these match the structures observed experimentally by Ziemann and Castleman[10]. Very recently Haertelt et al.[13] determined structures of neutral MgO clusters. The larger clusters obtained by our SA/GA+DFT technique is in a similar line with those obtained by Haertelt et al[13]. Only for $(MgO)_{12}$ the stucture of D_{3d} symmetry is found to be of lowest energy while Haertelt et al.[13] claimed a T_h structure as the global minimum. Local structures of neutral $(MgO)_n$ clusters are also reported and the relative energies are shown in Fig.5.1, 5.2 and 5.3. Both SA and GA are proven global optimizers and can locate global minimum in most situations with accuracy. In the present study since SA and GA methods explore the same PES (given in eqn. 1) the same global minimum was reached by both methods. The global minimum is a property of the PES and it is for the optimizer to hit that point on the surface in an optimization procedure. In our case both SA and GA were able to locate the same structure for each sizes of the cluster studied. SA starts with a single-trial solution and improves it iteratively by following an annealing schedule. GA on the other hand starts with a pool of solution and the entire set is evolved. Therefore for a single computational step, GA is costlier than SA, but this deficiency to a great extent is overcome because of lesser number of steps GA takes to reach the solution. The lesser number of steps taken is because GA has a greater diversity of search space to exploit from the outset. In our studies on the whole, we have seen that as regards computational time, GA takes about 20% more time to reach the structure compared with SA for a similar sized structure, but from the point of view of the number of computational steps needed, SA on an average takes double the number of steps when compared with GA.

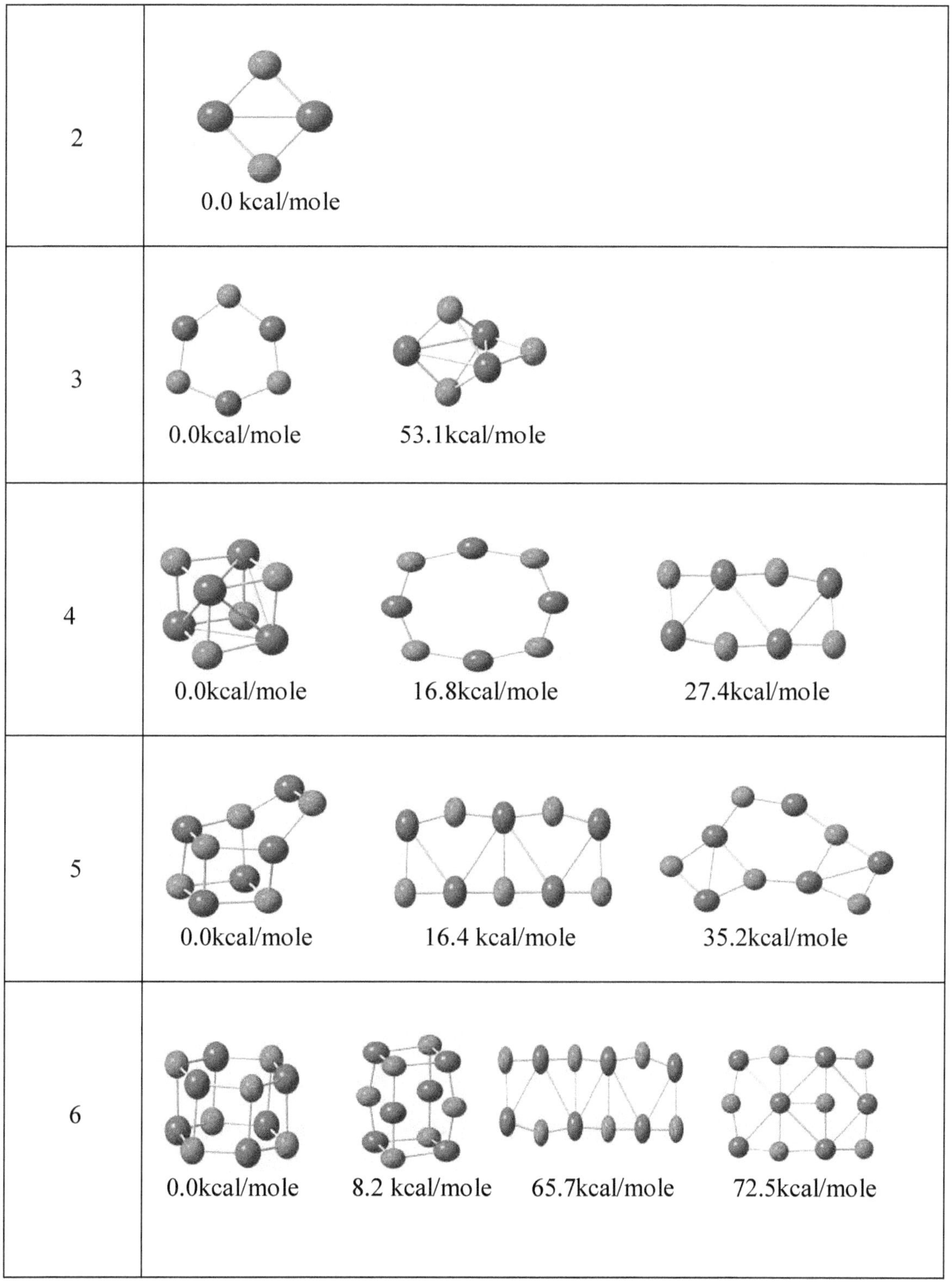

Fig. 5.1 Structures of (MgO)n clusters with n=2-6

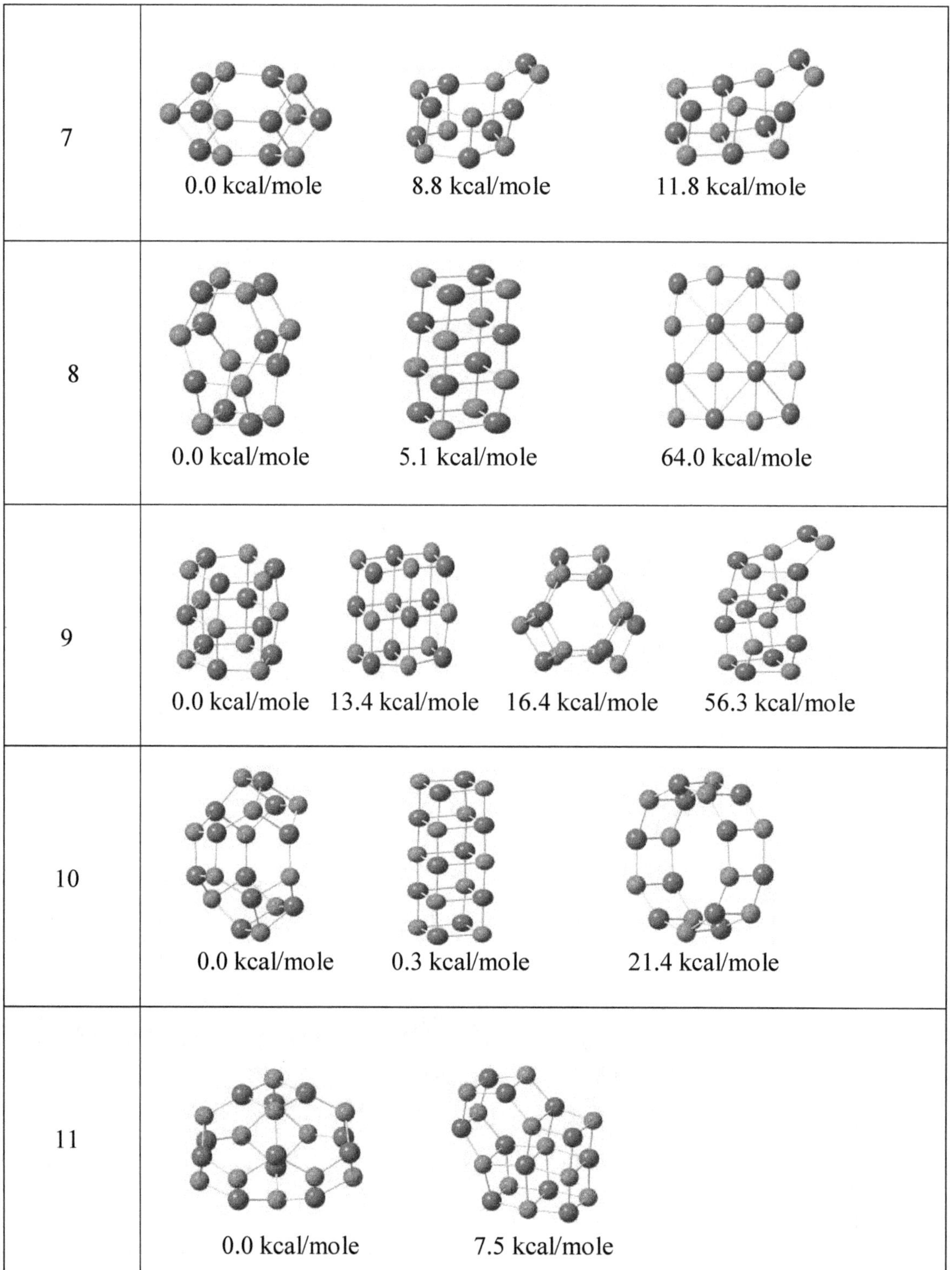

Fig. 5.2 Structures of $(MgO)n$ clusters with n=7-11

n	Structures
12	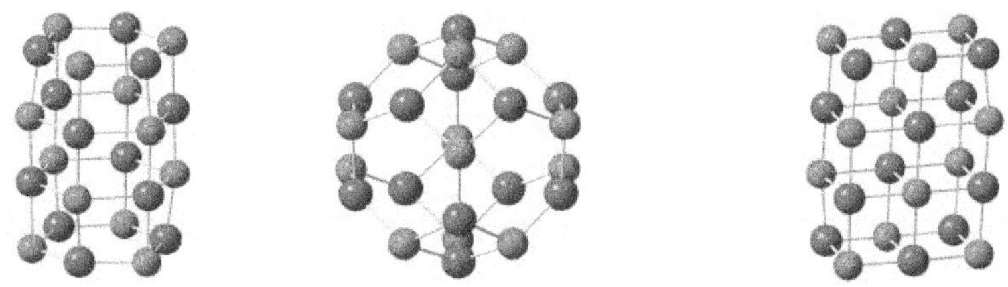 0.0 kcal/mole 3.5 kcal/mole 14.3 kcal/mole
13	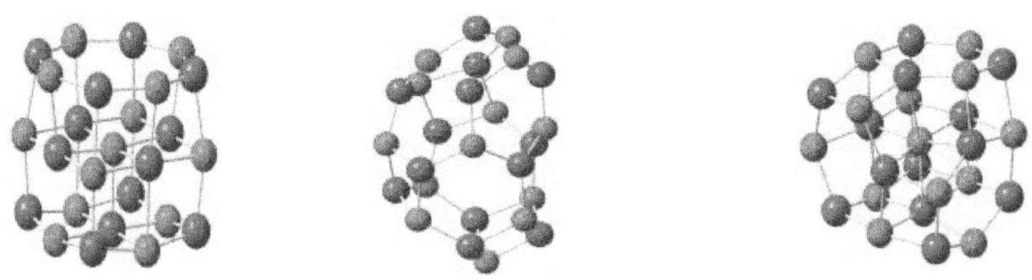 0.0 kcal/mole 4.9 kcal/mole 6.2 kcal/mole
14	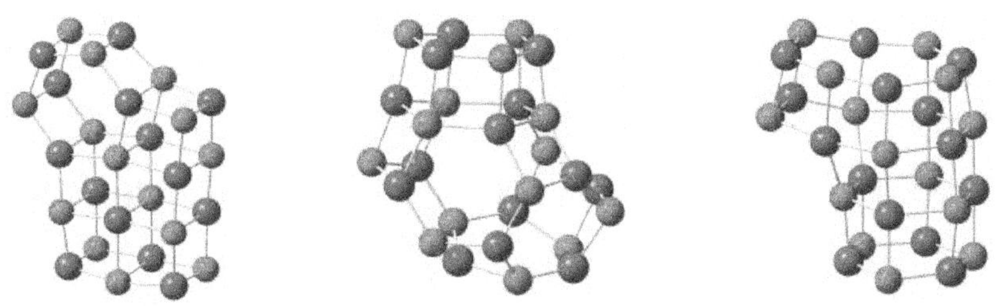 0.0 kcal/mole 4.6 kcal/mole 9.0 kcal/mole
15	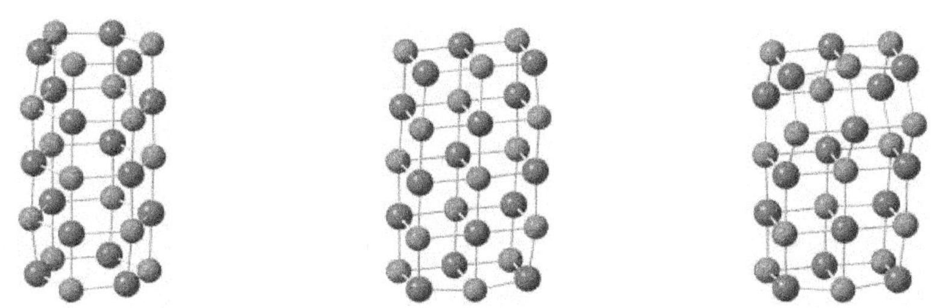 0.0 kcal/mole 15.0 kcal/mole 15.6 kcal/mole

Fig. 5.3 Structures of (MgO)n clusters with n=12-15

B. Spectroscopy of (MgO) n clusters

The infrared (IR) spectra of all the neutral global minimum (MgO)n clusters obtained at [B3LYP/6-311++G(d,p)] level are depicted in Fig. 5.4. Vibrational spectra of $(MgO)_n$ clusters are measured in the range 200-1000cm^{-1}, where characteristic vibrations of metal oxide clusters are expected to occur. The nature of binding of the dioxygen ligands can be identified by their infrared (IR) spectra. The superoxo species absorbs at 1100cm^{-1} while physisorbed oxygen molecule shows absorption near 1550cm^{-1}. Peroxo species shows characteristic vibration around 700-800cm^{-1}. Neither of the magnesium oxide cluster shows infrared absorption peak above 1000cm^{-1}, which signatures the absence of physisorbed oxygen molecule as well as superoxo species. We report here infrared (IR) spectra of MgO clusters with n=2-15. A low number of bands have been found for several sizes which indicate somewhat symmetric structure. Experimental results shows peaks for $(MgO)_3$ at 770cm^{-1} and 300cm^{-1}. The computed frequencies of $(MgO)_3$ are obtained at 758cm^{-1} and 268cm^{-1}. The structure of $(MgO)_4$ is a distorted cube which shows absorption signals at532cm^{-1}, 566cm^{-1} and 567cm^{-1} which are in good agreement to the experimentally observed signals. The bands experimentally observed at 280cm^{-1} and 820cm^{-1} can be explained by local octagonal isomer. The global minimum of $(MgO)_5$ shows an open book structure with characteristic IR active modes at 540cm^{-1}, 554cm^{-1}, 558cm^{-1}, 588cm^{-1}, 679cm^{-1} and 768cm^{-1}. The experimental bands observed at 560cm^{-1}, 690cm^{-1} and 775cm^{-1} are explainable by this hexagonal-cubic motif of $(MgO)_5$. The lowest energy (D_{3d}) structure of $(MgO)_6$ formed by two staggered $(MgO)_3$ show IR active modes at 367cm^{-1}, 368cm^{-1}, 526cm^{-1}, 670cm^{-1}, 677cm^{-1} and 678cm^{-1}. The calculated spectrum is in accordance with the experimentally observed peaks at 385cm^{-1},550cm^{-1}and 680cm^{-1}.The lowest energy isomer of $(MgO)_7$ possess a cage like structure of (C_{3v}) symmetry, with Mg and O atoms located evenly on the vertices. The cluster $(MgO)_7$ absorbs at 598cm^{-1}, 599cm^{-1} and 718cm^{-1}. Experiment reveals absorption bands at 595cm^{-1} and 720cm^{-1}. The global minimum of $(MgO)_8$ corresponds to a cage structure. This structure shows signals at 550cm^{-1}, 641cm^{-1}, 729cm^{-1} and 747cm^{-1}. For this conformer experiment shows absorption bands at 550cm^{-1}, 630cm^{-1} and 725cm^{-1}. The structure of $(MgO)_9$ corresponds to three stacked hexagonal rings. Infrared (IR) absorption peaks at 392cm^{-1}, 549cm^{-1}, 674cm^{-1} and 695cm^{-1} conforms to this structure. The global minimum of $(MgO)_{10}$ represents a cage structure of (C_2) symmetry with characteristic infrared absorption at 557cm^{-1}, 674cm^{-1} and 755cm^{-1}. The experimental signals are noticed at 550cm^{-1},685cm^{-1}and 740cm^{-1}. The number of bands in the vibrational spectra tends to increase with the increase in cluster size, with the consequent formation of broad absorption bands. As the cluster size increases, potential energy surface becomes complicated and numerous isomers lie close in energy. The hendecamer $(MgO)_{11}$ displays a fused cuboidal-hexagonal structure. For this geometry DFT calculation reveals infrared (IR) signals at 563cm^{-1}, 670cm^{-1}, 694cm^{-1} and 722cm^{-1}. The global minimum structure of $(MgO)_{12}$ is a hexagonal tube which displays a series of signals in the range 200-1000cm^{-1}. DFT calculation shows absorption bands at 486cm^{-1}, 657cm^{-1}, 675cm^{-1}, 678cm^{-1} and 705cm^{-1}. Experiment shows a broad absorption band around 700cm^{-1}. The global minimum of $(MgO)_{13}$ is an O-centered $3\times3\times3$ cube, with a Mg ion missing from the centre of one of the 3×3 cube faces, which then distorts to form an octagon. This structural motif displays IR active modes at 392cm^{-1}, 410cm^{-1}, 670cm^{-1}, 698cm^{-1}, 699cm^{-1}, 746cm^{-1} and 747cm^{-1}. The agreement of this spectrum to the experimental spectrum is very good. The global minimum structure of $(MgO)_{14}$ is quite different and contains both the cubic as well as the hexagonal ring motifs. The geometry displays vibrational bands at 570cm^{-1}, 584cm^{-1}, 700cm^{-1}, 729cm^{-1}, 738cm^{-1},748cm^{-1} and 768cm^{-1}. The lowest energy isomer of $(MgO)_{15}$ corresponds to a hexagonal tube of $(MgO)_3$ units which shows characteristic absorption bands at 647cm^{-1}, 651cm^{-1}, 676cm^{-1} and 682cm^{-1}. Our results are in good agreement to the experimental peaks obtained by Haertelt et al.[13].

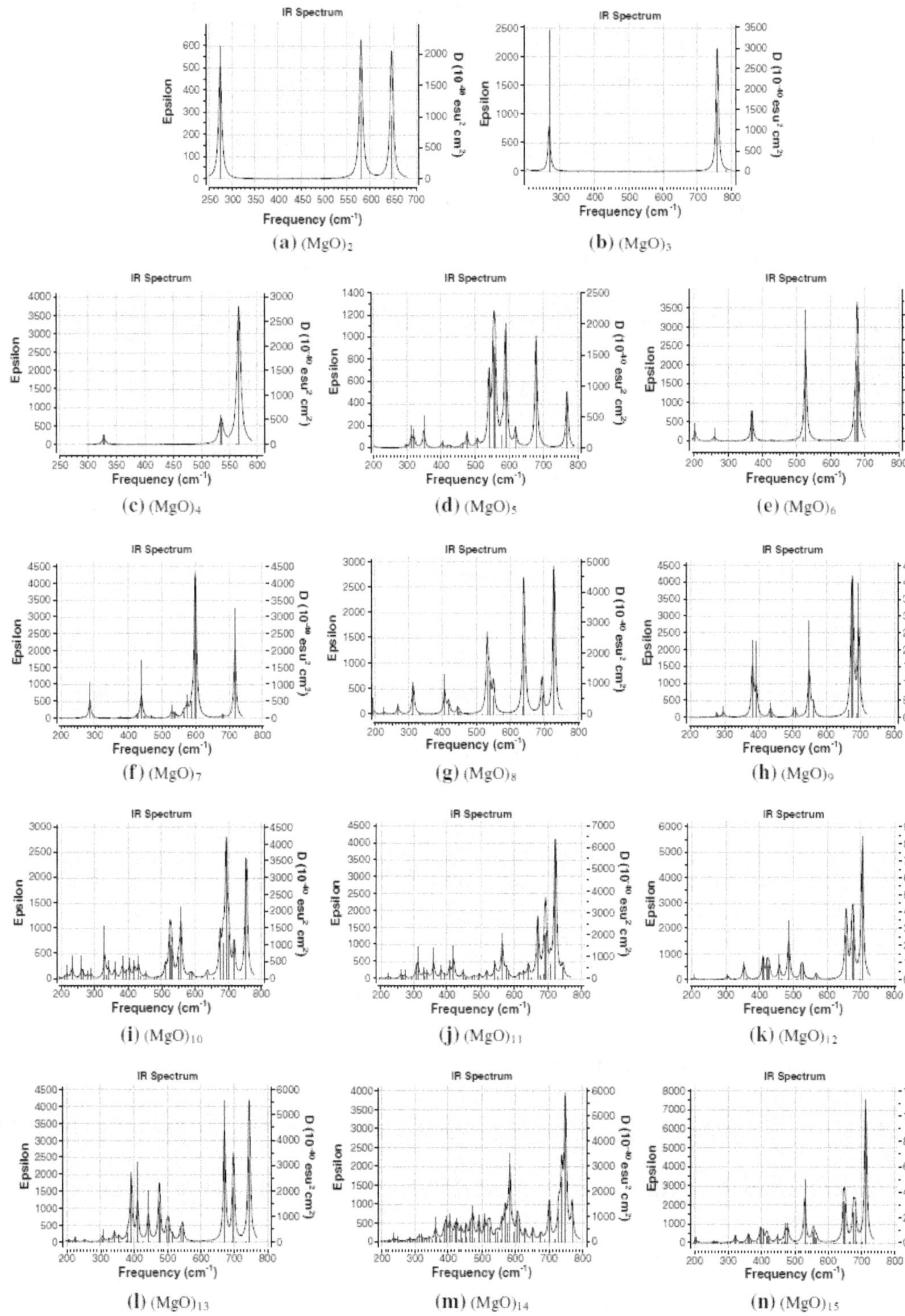

Fig. 5.4 Infrared spectra of (MgO)n clusters with n=2-15

C. Electronic Properties of (MgO)n clusters

For $(MgO)_n$ clusters, we have reported cluster-formation energies (E_{cf}), HOMO-LUMO gap(E_g), vertical ionization potential (IP_v), Singlet-Triplet gap (E_{st}) and vertical excitation energy (E_{ve}) in Table 5.3 [page 115]. The variations of these quantities are depicted pictorially in Fig. 5.5. It is obvious from Fig. 5.5 that the cluster formation energy increases for the smaller clusters and for the larger ones it shows constancy as expected. The (E_{cf}) and the (E_g) for the most stable lowest energy structure of $(MgO)_n$ clusters for each value of n have also been calculated. From the figure it is clear that at the beginning E_g increases and the plot show a dip at n=5, then the plot show an odd-even oscillation with change of cluster size. In general there is a direct relationship between the stability and E_g of the clusters, a larger E_g implies a higher stability for a particular system. The highest value of E_g is found for $(MgO)_{15}$ cluster system, which has been found to be the most stable structure. The vertical ionization potential (IP_v) decreases as the cluster size grows. An interesting odd-even oscillation is observed in the plot. This suggests that knocking of electron is easier for clusters of larger size. The vertical excitation energy (E_{ve}) i.e, the energy required in going from ground singlet to 1st excited singlet shows an increasing trend for the smaller size clusters, shows a dip for n=5, then shows an odd-even oscillation. The rate of oscillation gradually diminishes for $(MgO)_n$ clusters with increase of cluster size. In case of metal-oxide clusters, the E_g is often related to the stability. The cluster showing large Eg indicates a stable compound while those with lower E_g values shows the sign of greater reactivity. For $(MgO)_n$ clusters the HOMO is distributed mainly about oxygen anions and the LUMO about the magnesium atoms. The E_g of $(MgO)_6$ and $(MgO)_9$ are obviously larger for smaller clusters which demonstrates their higher stability. With the increase of size of $(MgO)_n$ clusters the average atomic charges increases. $(MgO)_n$ clusters with n=2-15 displays mixed covalent and ionic bonding. The variation of vertical ionization potential (IP_v) with cluster size(n) is displayed in Fig. 5.5. The plot shows dip at n=5, 7, 10, 13 cluster sizes. This inflexion can be explained on the basis of the fact that in these isomers there is always an oxygen anion, with lower number of neighbouring ions, which allows for knocking an electron more easily compared with other clusters, to form a

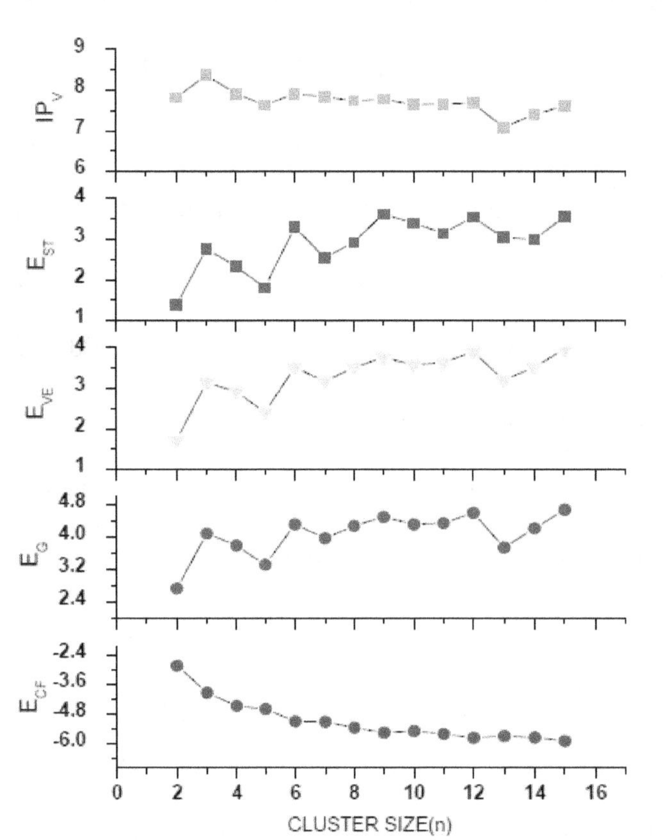

Fig.5.5 variation of electronic properties of (MgO)n clusters with n=2-15

single ionized cluster. The cluster-formation energies (E_{cf}) suggests that structural stability increases gradually with increasing length of MgO nanotube clusters. The variation of second difference (Δ^2E) in total energies of the clusters with cluster size (n) is depicted in Table 5.3 and it has been calculated using the relation [$\Delta^2E=2E_n-(E_{n+1}+E_{n-1})$] where En is the total energy of the (MgO)n cluster. Table 5.3 lists the Δ^2E values for the $(MgO)_n$ clusters. Δ^2E reflects the relative stability of the studied clusters (n) with respect to its neighbours (n-1 and n+1). Positive (negative) Δ^2E means unstable (stable) clusters with respect to its neighbours. The plot of Δ^2E vs n(Fig. 5.6) shows a clear odd-even oscillation with increasing n. Δ^2E is negative for cluster sizes with n=4, 6, 9 and 12. From figure 5.6 we observe that the clusters $(MgO)_4$, $(MgO)_6$,

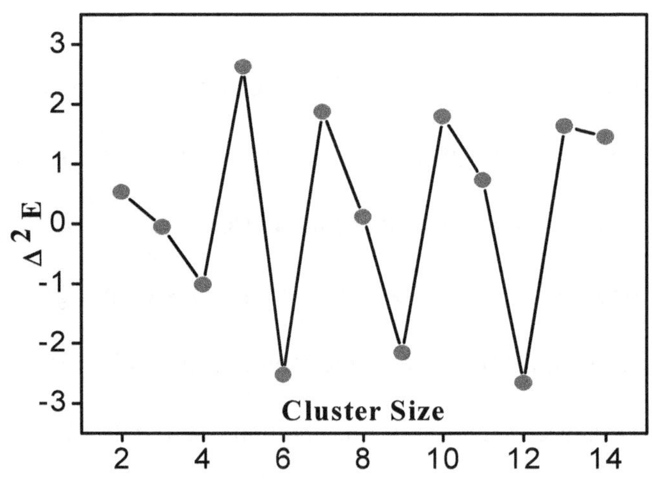

Fig. 5.6 variation of seconed-order difference of total energy of (MgO)n clusters with n=2—15

$(MgO)_9$ and $(MgO)_{12}$ are more stable in comparison to their neighbours. The $(MgO)_{12}$ cluster has the lowest value of Δ^2E which is attributed to its highest stability.

Table 5.3. B3LYP/6-311++G(d,p) calculated cluster-formation energies per MgO unit (with zpe correction), HOMO-LUMO gaps (E_g), vertical ionization potential (IPv) and TDDFT/B3LYP Vertical Excitation Energies(E_{ve}), singlet-triplet gaps (E_{st}), second-order difference of total energy (Δ^2E) for global minimum $(MgO)_n$ clusters in eV units.

N	E_{cf}	Δ^2E	E_{homo}	E_{lumo}	E_g	E_{ve}	E_{st}	IP_v	Lit[13]
2	-2.82	0.53	-5.64	-2.90	2.74	1.72	1.38	7.82	7.71
3	-3.94	-0.06	-6.38	-2.28	4.10	3.15	2.76	8.37	8.49
4	-4.47	-1.02	-6.04	-2.25	3.79	2.92	2.33	7.92	8.27
5	-4.60	2.62	-5.78	-2.46	3.32	2.41	1.81	7.64	7.56
6	-5.10	-2.53	-6.25	-1.93	4.32	3.52	3.30	7.91	7.87
7	-5.12	1.87	-6.07	-2.10	3.97	3.18	2.55	7.84	8.04
8	-5.36	0.11	-6.21	-1.94	4.28	3.51	2.93	7.74	7.85
9	-5.56	-2.16	-6.30	-1.80	4.50	3.76	3.62	7.78	7.94
10	-5.50	1.79	-6.22	-1.90	4.32	3.58	3.40	7.65	7.62
11	-5.62	0.73	-6.24	-1.89	4.35	3.63	3.15	7.66	7.64
12	-5.77	-2.66	-6.35	-1.75	4.60	3.90	3.54	7.69	7.75
13	-5.70	1.63	-5.63	-1.89	3.74	3.21	3.05	7.08	7.27
14	-5.76	1.45	-6.12	-1.90	4.22	3.52	3.00	7.41	7.54
15	-5.90		-6.39	-1.71	4.68	3.97	3.56	7.62	7.70

D. Polarizability and First Hyperpolarizability of (MgO)n clusters

Once the structure and the IR spectra of the $(MgO)_n$ cluster systems are obtained using our hybrid technique of combining stochastic optimization technique with DFT calculations we proceed on to compute the polarizability(α) and first hyper-polarizability(β) of all the lowest minimum energy structures. The results are summarized in Table 5.4. It is well known that DFT method generally predicts lower values for polarizability and higher for first order hyperpolarizability when compared with other techniques like MP2.

Table 5.4 B3LYP/6-311++G(d,p) calculated static and dynamic polarizability (α) and first hyperpolarizability (β) for the global minimum of (MgO)n clusters in a.u. The incident field frequency is ω=0.0428 a.u.

N	α(0;0) (static)	α(-ω;ω) (dynamic)	β (0;0,0) (static)	β (-2ω; 0,0) (dynamic)
2	59.88	66.92	0.23	0.31
3	74.51	77.65	6.25	17.6
4	88.16	92.05	7.41	18.04
5	112.76	117.78	2903.27	2118.2
6	121.07	124.52	0.07	0.12
7	142.23	146.41	1869.79	3174.4
8	157.45	161.49	0.0003	0.0006
9	170.84	174.98	8.98	25.4
10	197.96	202.53	305.41	504.90
11	210.53	215.35	598.75	884.30
12	221.03	225.98	4.82	7.30
13	243.10	249.41	5201.46	10761.10
14	259.36	265.50	647.67	1197.77
15	271.37	277.17	66.63	107.31

Still the values are a good measure for following trends with increasing size for a given cluster systems as well as a confirmatory check on the symmetry of the computed structures especially in situations where the structures possess a centre of symmetry. The mean dipole polarizability of a system is given by

$$\alpha = \frac{1}{3}\left(\alpha_{xx} + \alpha_{yy} + \alpha_{zz}\right)$$

Where α_{xx}, α_{yy} and α_{zz} are the polarizability tensor components.

The first hyperpolarizability can be written as

$$\beta = (\beta_x^2 + \beta_y^2 + \beta_z^2) \text{ Where}$$

$$\beta_x = \beta_{xxx} + \beta_{xyy} + \beta_{xzz}$$

$$\beta_y = \beta_{yxx} + \beta_{yyy} + \beta_{yzz}$$

$$\beta_z = \beta_{zxx} + \beta_{zyy} + \beta_{zzz}$$

and $\beta_{xxx}, \beta_{xyy}, \beta_{xzz}$,.....etc. are the usual first hyperpolarizability components.

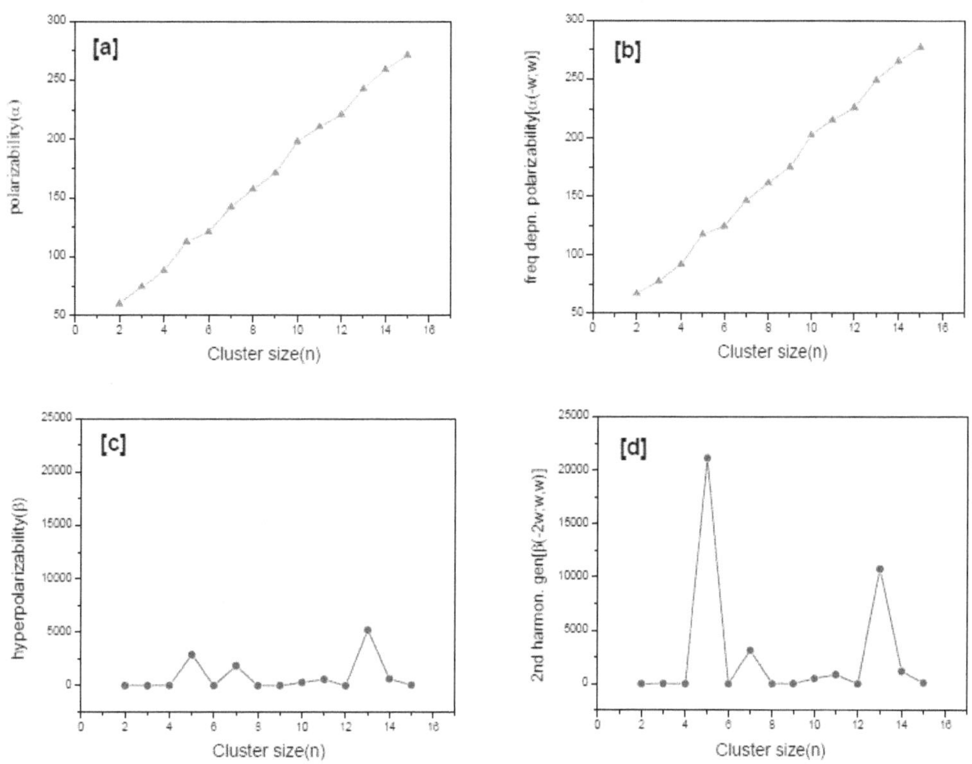

Fig. 5.7 Variation of nonlinear optical properties of (MgO)n clusters with n=2-15

If we look at the Table 5.4 and Fig. 5.7 we see that the polarizability increases with the size of the cluster systems, both in the presence and absence of field having a particular frequency [here the frequency source is the Nd:Yag laser with frequency 0.0428 a.u. (λ=1064 nm)]. This is an expected trend shown by most systems that with the increase in size the polarisable electrons are less tightly bound with the nuclear centres and hence are more polarizable. However the first hyperpolarizability variation with size is more interesting with oscillatory patterns and specially for some sizes of $(MgO)_n$ the value being close to zero. This can be seen clearly in Fig. 5.7. The zero first hyperpolarizability scenario is observed for n=2, 3, 6, 9, 12 etc. This can be explained based on the fact that these clusters are highly symmetric with a centre of symmetry. The figures showing the clusters are a testimony to this. For the centre of symmetry possessing systems both the first hyper polarizability as well as dipole moment vanishes to zero. The vanishing of the first hyperpolarizability is a further confirmatory conclusion regarding the high symmetry of some of the sizes of $(MgO)_n$ system.

5.4 Conclusion

A hybrid SA/GA-DFT method is employed to generate the lowest energy isomers of $(MgO)_n$ clusters with n=2-15. We have studied the structures, spectroscopy, growth behaviour and electronic properties of neutral $(MgO)_n$ clusters with n=2-15. The interconnection between electronic and structural properties has been studied by considering the variations of cluster-formation energies (E_{cf}), HOMO-LUMO gap (Eg), vertical ionization potential (IP_v) and vertical excitation energy (E_{ve}) with cluster-size(n). The combination of Density Functional Theory (DFT) in conjunction with Simulated Annealing (SA)/Genetic Algorithm (GA) in association with infrared (IR) spectroscopy establishes that when n is a multiple of 3, hexagonal tube is the most stable structural motif. For the global minimum $(MgO)_n$ clusters, we have evaluated the HOMO-LUMO gaps (Eg), vertical ionization potential (IP_v), vertical excitation energy (E_{ve}), cluster-formation energies (E_{cf}) and second-order difference of total energy (Δ^2E). From the second-order difference of total energy (Δ^2E) we find that the n=5 & 7 cluster is thermodynamically unstable with respect to disproportionation to smaller and larger clusters. Δ^2E versus cluster size (n) plot shows that $(MgO)_{3n}$ clusters are more stable. We find extreme quicker convergence using SA/GA+DFT technique. There is a very close correspondence of the computed infrared modes to the experimental frequencies. Since SA/GA converges very quickly to a structure which is very near to a minimum obtained from a quantum chemical calculation, hence if SA/GA structures are used as inputs in quantum chemical calculations convergence is seen to be extremely quick as compared to a pure quantum chemical calculation.

5.5 References

[1] S. K. Sahoo, S. Pal, P. Sarkar and C. Majumder Chem. Phys. Lett **2011**, 516, 6871

[2] S. Saha, S. Sarkar, S. Pal and P. Sarkar RSC Adv. **2013**, 3, 532

[3] S. Pal, R. Sharma, B. Goswami and P. Sarkar Chem. Phys. Lett **2009**, 467, 365

[4] S. H. C. Liang, I. D. Gay J. Catal **1986**, 101, 293

[5] H. Tsuji, F. Yagi, H. Hattori, H. Kita J. Catal **1994**, 148, 759

[6] G. Bilalbegovic Phys. Rev. B **2004**, 70, 045407-1

[7] R. Dong, X. Chen, X. Wang, W. Lu J. Chem. Phys **2008**, 129, 044705

[8] Z. Cui, G W Meng, W. D. Huang, G. Z. Wang, L. D. Zhang Matter. Res. Bull **2000**, 35, 1653

[9] Q. Yang, J. Sha, L. Wang, Y. Wang, X. Ma, J. Wang and D. Yang Nanotechnology **2004**, 15, 1004

[10] P. J. Ziemann, A. W. Castleman J. Chem. Phys **1991**, 94, 718

[11] W. A. Saunders Z. Phys. D **1989**, 12, 601

[12] W. A. Saunders Phys. Rev. B **1998**, 37, 6583

[13] M. Haertelt, A. Fielicke, G. Meijer, K. Kwapien, M. Sierka, J. Sauer Phys. Chem. Chem. Phys **2012**, 14, 2849

[14] S. Moukouri, C. Nogurea Z. Phys. D **1992**, 24, 71

[15] J. M. Recio, R. Pandey Phys. Rev. A **1993**, 47, 2075

[16] J. M. Recio, R. Pandey, A. Ayuela, A. B. Kunz J. Chem. Phys **1993**, 98, 4783

[17] A. Jain, V. Kumar, M. Sluiter, Y. Kawaoze Comput. Mater. Sci **2006**, 36, 171

[18] L. Hong, H. Wang, J. Cheng, L. Tang, J. Zhao, J. Theo. Comput. Chem **2012**, 980,62

[19] M. J. Malliavin, C. J. Coudray, Chem. Phys., **1997**, 106, 2323

[20] E. de. la. Puente, A. Aguado, A. Ayuela, J. M. Lopez, Phys. Rev. B. Condens. Mat. **1997**, 56, 7607

[21] T. M. Kohler, H. P. Gail, E. Sedlmayr, Astron. Astrophys. **1997**, 320, 553

[22] M. Wilson, J. Phys. Chem. B., **1997**, 101, 4917

[23] F. Calvo, Phys. Rev. B. Condens. Mat. **2003**, 67, 161403

[24] J. S. Bhatt, I. J. Ford, Mon. Not. R. Astron. Soc **2007**, 382, 291

[25] C. Roberts, R. L. Johnston, Phys. Chem. Chem. Phys. **2001**, 3, 5024

[26] J. Carrasco, F. Illas, S. T. Bromley, Phys. Rev. Lett. **2007**, 99, 235502

[27] Y. Zhang, H. S. Chen, B. X. Liu, C. R. Zhang, X. F. Li, Y. C. Wang J. Chem. Phys., **2010**, 132, 194304

[28] L. Chen, C. Xu, X. F. Zhang, Theochem **2008**, 863, 55

[29] L. Chen, C. Xu, X. F. Zhang, C. A. Cheng, T. Zhou, Int. J. Quant. Chem. **2009,** 109,349

[30] L. Andrews, J. T. Yustein, J. Chem. Phys. **1993**, 97, 12700

[31] D. Van. Heijnsbergen, G. Von. Helden, G. Meijer, M. A. Duncan, J. Chem. Phys. **2002,** 116, 2400

[32] K. Kwapien, M. Sierka, J. Dobler, J. Sauer, M. Haertelt, A. Fielicke and G. Meijer, Angew. Chem. Int. Edn **2011**, 50, 1716

[33] S. Sicolo, L. Giordano, G. Pacchioni J. Phys. Chem. C 2009**,** 113, 16694

[34] M. Pesci, F. Gallino, C. D. Valentin, G. Pacchioni J. Phys. Chem. C 2010**,** 114, 1350

[35] A. Roldan, J. M. Ricart, F. Illas, G. Pacchioni J. Phys. Chem. C **2010,** 114, 16973

[36] G. Pacchioni, H. Freund Chem. Rev **2013,** 113, 4035

[37] J. P. K. Doye Phys. Rev. B. **2003,** 68, 195418

[38] S. M. Woodley, C. R. A Catlow, Nature Mater. **2008**, 7, 937.

[39] S. M. Woodley, S. Hamad, C. R. A. Catlow, Phys. Chem. Chem. Phys. **2010**,12, 8454.

[40] C. R. A. Catlow, S. T. Bromley, S. Hamad, M. Mora-Fonz, A. A. Sokol, S. M. Woodley Phys.Chem. Chem. Phys. **2010**, 12, 786

[41] Aron Walsh, S. M. Woodley, Phys. Chem. Chem. Phys. **2010**, 12, 84468453

[42] W. Q. Huang, X. J. Lai, R. C. Xu, Chem. Phys. Lett. **2011**, 507, 199.

[43] X. J. Lai, W. Q. Huang, R. C. Xu, J. Phys. Chem. A. **2011**, 115, 5021.

[44] S. Kirkpatrick, C. D. Gelatt, M. P. Vecchi, Science **1983**, 220, 671

[45] S. Kirkpatrick, J. Stat. Phys. **1984**, 34, 975

[46] D. E. Goldberg, Genetic Algorithm in Search, Optimization and Machine learning (Addison Wesley: Reading MA, 1989)

[47] J. Holland, Adaptation in Natural and Artifcial Systems (MIT Press, 1992)

[48] M. Mitchell, An introduction to Genetic Algorithm (MIT Press, Cambridge, MA, 1998)

[49] D. J. Earl, M. W. Deem, Phys. Chem. Chem. Phys **2005**, 7, 3910

[50] J. P. K. Doye, D. J. Wales, J. Phys. Chem. A. **1997,** 101, 5111

[51] R. H. Leary, J. P. K. Doye, Phys. Rev. E. **1999**, 60, R6320.

[52] R. H. Leary, J. Global Optim. **2000,** 18, 367

[53] D. J. Wales. H. A. Scheraga, Science **1999**, 285, 1368

[54] J. Hernandez-Rojas, J. Breton, J. M. Gomez Llorente, D. J. Wales, Chem. Phys. Lett **2005**, 410, 404

[55] B. S. Gonzalez, J. Hernandez-Rojas, D. J. Wales Chem. Phys. Lett **2005**, 412, 23

[56] J. Hernandez-Rojas, D. J. Wales, J. Chem. Phys **2003**, 119, 7800

[57] J. Hernandez-Rojas, J. Breton, J. M. Gomez Llorente, D. J. Wales J. Chem. Phys **2004**, 121, 12315

[58] X. J. Lai, W. Q. Huang, R. C. Xu, J. Phys. Chem. A **2011**, 115, 5021

[59] X. G. Shao, L. J. Cheng, W. S. Cai, J. Comput. Chem. **2004**, 25, 1693

[60] X. L. Yang, W. S. Cai, X. G. Shao, J. Comput. Chem. **2007**, 28, 1427

[61] X. G. Shao, X. L. Yang, W. S. Cai, J. Comput. Chem. **2008**, 29, 1772

[62] X. Wu, W. S. Cai, X. G. Shao, Chem. Phys. **2009**, 363, 72

[63] R. L. Johnston, Dalton Trans. **2003**, 22, 4193

[64] L. Davis, Handbook of Genetic Algorithms; Van Nostrand Reinhold: New York,**1991**.

[65] J. R. Morris, D. M. Deaven, K. M. Ho, Phys. Rev. B **1996**, 53, 1740

[66] B. Hartke, J. Phys. Chem. **1993,** 97, 9973

[67] N. O. Carstensen, J. M. Dieterich and B. Hartke, Phys. Chem. Chem. Phys 2011, 13, 2903-2910

[68] J. M. Dieterich, B. Hartke, J. Comput. Chem **2011**, 32, 1377-1385

[69] Y. Xiao, D. E. Williams, Chem. Phys. Lett. **1993**, 215, 17

[70] Y. Zeiri, Phys. Rev. E. **1995**, 51, 2769

[71] Y. Zeiri, J. Phys. Chem. A. **1998**, 102, 2785.

[72] D. X. Tian, H. L. Zhang, J. J. Zhao, Solid. State. Commun. **2007**, 144, 174.

[73] K. Oleksy, F. Karlicky, R. Kalus, J. Chem. Phys., **2010**, 133, 164314.

[74] J. L. Llanio-Trujillo, J. M. C. Marques, F. B. Pereira, J. Phys. Chem. A **2011**, 115, 2130.

[75] D. M. Deaven, K. M. Ho, Phys. Rev. Lett. **1995**, 75, 288

[76] D. M. Deaven, J. R. Morris, K. M. Ho, N. Tit, Chem. Phys. Lett. **1996**, 256 ,195

[77] W. J. Pullan, J. Chem. Inf. Modelling **1997**,37, 1189

[78] W. J. Pullan, Comput. Phys. Comm. **1997,** 107, 137

[79] W. J. Pullan, J. Comput. Chem. **1997**, 18, 1096

[80] J. A. Niesse, H. R. Mayne, Chem. Phys. Lett. **1996,** 261, 576

[81] J. A. Niesse, H. R. Mayne, J. Chem. Phys. **1996**, 105, 4700

[82] J. A. Niesse, H. R. Mayne, J. Comput. Chem. **1997**, 18, 1233

[83] R. L. Johnston, Dalton. Transc. **2003**, 4193

[84] C. Roberts, R. L. Johnston, N. T. Wilson, Theo. Chem. Acc. **2000**, 104, 123

[85] R. Ismail, R. L. Johnston, Phys. Chem. Chem. Phys. **2010**, 12, 8607

[86] K. Michaelian, Chem. Phys. Lett. **1998**, 293, 202

[87] K. Michaelian, Am. J. Phys. **1998**, 66 , 231

[88] K. Michaelian, I. L. Garzon, M. R. Beltran, P. Ordejon, D. Portal-Sanchez, J. M. Soler, E. Artacho, Phys. Rev. Lett. **1998**, 81, 1600

[89] S. G. Neogi, P. Chaudhury, J. Comput. Chem. **2012**, 33, 629

[90] S. G. Neogi, P. Chaudhury, J. Comput. Chem. **2013**, 34, 471

[91] B. Hartke, Chem. Phys. Lett. **1995**, 240, 560

[92] B. Hartke, Chem. Phys. Lett. **1996**, 258, 144

[93] S. K. Gregurick, M. H. Alexander, B. Hartke J. Chem. Phys., **1996**, 104, 2684

[94] B. Hartke, M. Schutz, H. J. Werner, Chem. Phys. **1998**, 239, 561

[95] Gaussian 09, Revision B.01, M. J. Frisch

Chapter 6

Structural, Spectroscopic Aspects, and Electronic Properties of (TiO_2)n Clusters: A study Based on the Use of Natural Algorithms in Association with Quantum Chemical Methods

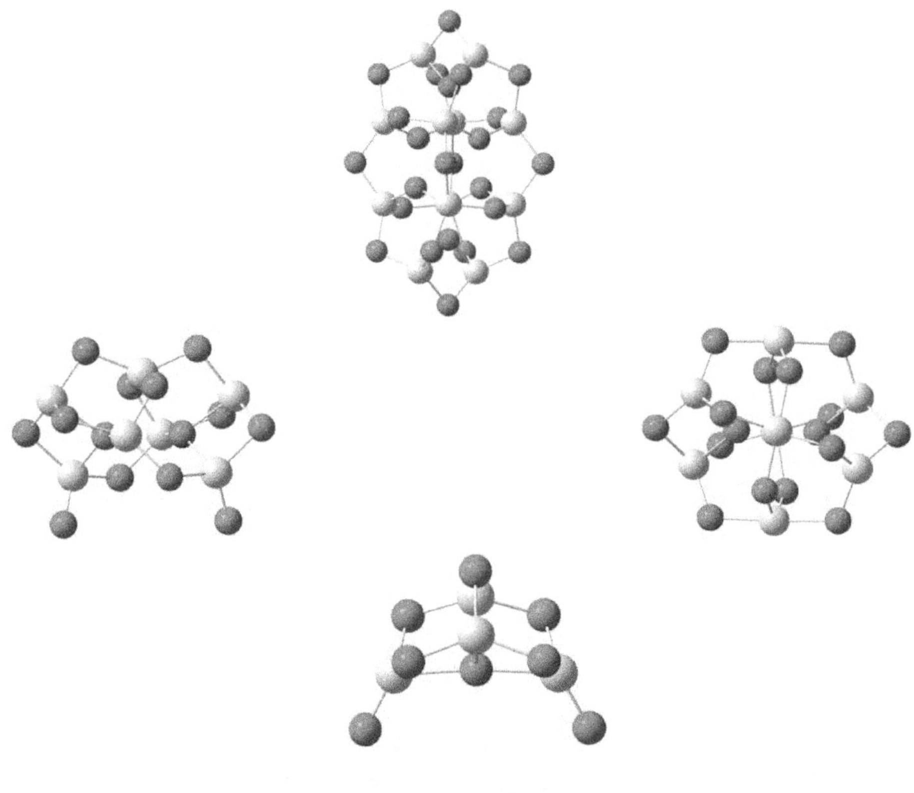

S.G. Neogi, P. Chaudhury J. Comput. Chem. 2014, 35, 51-61

6.1 Introduction

The study of atomic and molecular clusters is an important area of research for both experimentalists and theoreticians in the last few decades. These systems are interesting from the points of view of enormous structural diversity as well as the spectroscopic signatures that these subtle variations in structural forms contribute. Transition metal oxide clusters [1–13] have attracted the attention of researchers due to their widespread application in catalysis. Structure and electronic properties of TiO_2 nanoparticles is an active field of research owing to its application in photocatalysis and photovolatics. Studies on clusters of particles in the nanometer range have revealed some unusual bonding geometries. An ideal model compound for the study of metal-oxide nanoparticles is the titanium oxide because of the abundance of literature on this compound. Clusters of titanium oxide are very important for their widespread application [14–23] in diverse areas of chemistry, physics and material science. Tatiana (TiO_2) is the fundamental material in the field of gas-sensors, photovolatics, photocatalysis, and energy storage. The semiconductor property of titania (TiO_2) find its important activity in solarcells[24], photocatalysis[25,26], bacterial inactivation[27], water disinfection[28] and sunscreens[29]. Because of its chemical and biological [20] inertness as well as being inexpensive titania (TiO_2) is widely used as a white pigment. There has been considerable work on TiO_2 nanostructures due to its significant application as photocatalyst for hydrogen production from water using solar energy and for removal of organic pollutants in water or air[14, 15, 19]. Very recently Fang and Dixon[30] studied H_2 and O_2 production from water splitting by small (MO_2)n clusters (M=Ti,Zr,Hf; n=1-3). It is essential to explore the structure and stabilities of nanoparticles for investigation of quantum size effects on the photochemical properties. Subtle variations in physical and chemical properties with the change in cluster size provide significant insights into the growth behaviour of TiO_2 nanostructures. There has been considerable work on experimental [31–39] and theoretical [40–51] studies of titania (TiO_2) nanoclusters, especially the study regarding hydrolysis reactions in the ground and first excited triplet states by Wang et al.[51]. Tsipis et al.[52] first characterized TiO_2 cluster model. Collison-induced dissociation of cationic TiO_2 clusters were studied by Yu and Freas[33] using mass spectrometry. Chemical reaction of titanium oxide cluster cations with oxygen was studied by Guo et al.[34]. Single photon ionization mass spectra of Ti_mO_n neutral clusters were investigated by Matsuda and Bernstein[38]. Anion photoelectron spectra of $(TiO_2)_n^-$ (n=2-10) clusters were measured by Zhai and Wang[53]. Wei et al.[54] studied magnetism of anatase clusters experimentally. The structures of TiO_2 nanoclusters have been studied by several research groups theoretically. Ramana and Philips[40] studied ground state and low lying excited state of TiO_2 using configuration interaction calculation. Yu and Freas [33] found structures of (TiO_2)n neutral clusters for n=2-4 involving a simple pair potential model. Hamad et al.[46] reported various structures of TiO_2 clusters using pair potential based simulated annealing Monte Carlo basin hopping (SA-MCBH) and genetic algorithm(GA) methods. Coupled-cluster calculations involving single and double excitations was used by Li and Dixon[6]. Density functional calculations on (TiO_2)n clusters were performed by Shevlin et al.[55], Woodley et al.[45]. Qu and kroes[47,48] studied TiO_2 clusters using B3LYP/LANL2DZ method and obtained lower energy structures. Structure and reactiviy of small size $(TiO_2)_{1-10}$ has been investigated by Calatayud et al.[56,57]. Bandyopadhyay and Aikens [58] studied structure and stability of small (TiO_2)n clusters. Barnard et al.[59] have studied the structural and electronic properties of TiO_2 nanoparticles through self-consistent tight binding method and DFT. A topological method adopted by Tang et al.[60] in the global structure determination is worth mentioning. Hydrogenation of (TiO_2)n clusters(n=1-10) have been investigated by Syzgantseva et al[61]. Very recently, Cakir and Gulseren[62] have performed ab initio study of neutral (TiO_2)n clusters and their interactions with water and transition metal atoms. Chiodo et al.[63] investigated structural, electronic and optical properties of (TiO_2)n clusters using ab initio DFT calculation. Tsuchiya and Whitten[64] performed configuration interaction study on $(TiO_2)_{2n}$ clusters with n=3-9. Zhang et al.[65] performed stability analysis of titanium dioxide clusters with n=1-9. The study of (TiO_2)n clusters with n=1-4 involving simple pair-potential model[33] prompted ab initio and DFT studies for small neutral and charged (TiO_2)n clusters with n=1-3[41–44, 50, 66]. Jeong et al.[49] investigated structure and electronic properties of small (TiO_2)n clusters with n=1-6 by DFT calculation. Hagfeldt et al.[50] employed ab initio SCF methods and studied both the geometric and electronic structures of $(TiO_2)_2$ and $(TiO_2)_3$. Mowbray et al.[67] investigated stability and electronic properties of TiO_2 nanostructures with and without B and N doping. Very recently Rana et al.[68] performed ab initio study to understand stability and magnetism of (TiO_2)n clusters.

Despite the intensive studies discussed above, the most stable structures of (TiO_2)n clusters are still under controversy. A theoretical study of possible structures of these systems can present interesting challenges to optimization techniques and procedures. Deciphering the correct structure would essentially mean finding the correct set of interatomic distances for which the potential energy is a minimum. Deterministic or gradient based techniques work very well for single minimum problem, but for multiple minima systems, the search usually converges to a point where there is a quickest decrease of gradient. Hence the search becomes totally dependent on the initial point and the specificity of locating the global minimum is lost. In such situations, there has been increasing tendency to use truly global and stochastic optimization methods Simulated Annealing(SA) or Genetic Algorithm(GA). SA[69,70] borrows its concepts from metallurgical process of annealing. As in metallurgical annealing, there is a temperature like quantity which controls the thermal fluctuations while searching a rugged potential energy landscape. At high temperatures, the search, even if trapped in a local attractive basin, will be able to come out of it and surmount the barrier separating it from other minima and after having sampled the search space at different temperatures, will ultimately be able to unequivocally locate the global minimum in the limit of annealing temperature tending to zero. Genetic Algorithm (GA)[71-73] is another stochastic algorithm. GA draws its working philosophy from the genetic processes of selection, crossover and mutation. GA starts from a pool of trial solutions and improves it by above mentioned operations till the optimum desired solution is found out. GA has been used to solve a host of problems ranging from structure elucidation of atomic and molecular clusters, solution of schrodinger equation, predicting geometries of polymeric materials. The use of GA in elucidating structures of clusters has a rich and long history. Cluster systems ranging from Lennard-Jones cluster, Morse cluster, metallic clusters modelled by Gupta potential and work of several leading research groups needs mention. The work of Deaven et al.[74,75], Pullan et al.[76-78], Niesse and Mayne[79-81], Johnston et al.[82-84], Michaelian et al. [85-87], Ganguly Neogi et al.[88,89], Hartke et al.[90-93] deserves special attention. Recently there has been an impressive application by Chen and Dixon[94] using Tree Growth–Hybrid Genetic Algorithm along with quantum chemical calculations for predicting the structures of small (TiO_2)n clusters with n=2-13.

A new strategy of studying chemical systems is also becoming popular in which one uses GA/SA in conjunction with standard quantum chemical procedures, especially density functional theory to study large cluster systems. In this chapter a GA/SA + DFT method is employed for studying (TiO_2)n clusters with n=1-12 and once the structures are obtained we performed DFT calculations with B3LYP being the functional used and 6-31+G(d) being the basis set used taking the optimized structure from SA/GA as input.Various electronic properties of (TiO_2)n clusters are evaluated from DFT calculation. All DFT calculations have been performed using GAUSSIAN 09 [95] package.

6.2. The Methodology

A. The Potential for the system

In order to describe the interaction that exist between all the atoms constituting a cluster of the type $(TiO_2)n$, where n is the number of Ti atoms we have used the interatomic potential of Matsui and Akaogi[96]. The potential energy function consists of a two-body coulomb-plus-Buckingham potential. The total energy of the cluster is given by

$$E = \frac{1}{2}\sum\left[\frac{q_i q_j}{r_{ij}} + A_{ij}\, exp\left(\frac{-r_{ij}}{\rho_{ij}}\right) - \frac{C_{ij}}{r_{ij}^6}\right]$$

The parameters are given in Table 6.1. The charges are Ti=2.196|e| and O= -1.098|e| and the values of A, ρ and C are given in Table 6.1.

Table 6.1 Parameters used in potential energy function for $(TiO_2)n$

i—j	A_{ij} (kcal/mole)	ρ_{ij} (Å)	C_{ij}(Å^6)
Ti—O	391049.1	0.194	290.3317
O—O	217716.3	0.234	696.8883
Ti—Ti	717647.4	0.154	121.0676

B. Computational details

The problem we are studying in the present chapter is to get the right structure of the $(TiO_2)_n$ system, such that the potential energy of the system is a minimum. So essentially, the solution of the problem is to find optimum coordinates for all the atoms in the cluster. We have used Simulated Annealing and Genetic Algorithm in finding out the structures of $(TiO_2)n$ cluster systems. The details of Simulated Annealing and Genetic Algorithm have been described in Chapter 1 of the thesis. The Cartesian coordinates obtained from SA/GA simulation are used as input in geometry optimization using DFT. All DFT calculations are performed with GAUSSIAN 09 [95] package. We have used Becke's three-parameter exchange functional[97] along with Lee-Yang-Parr's correlation functional[98] i.e., the B3LYP functional, and the 6-31+G(d) basis set[99,100] as implemented in Gaussian09. The 6-31+G(d) basis set includes polarization and diffuse functions on Ti and O atom that is necessary for proper description of electronic structure of TiO_2 cluster system. The cartesian coordinates obtained from GA-SA simulation are used as input in geometry optimization using DFT. All neutral singlet (TiO_2)n clusters with n=1-12 are optimized and the absence of imaginary frequency indicates that these are truly minima on the potential energy hypersurface. Using B3LYP/6-31+G(d) [97-102] we have evaluate various properties of (TiO_2)n clusters e.g.; vibrational frequencies, cluster formation energies, vertical ionization potential(IP_v), vertical electron affinity(EA_v), singlet-triplet gaps, and so forth.

To compare the stability of (TiO_2)n particles with different n values we define the cluster formation energy per TiO_2 unit as follows:

$$E_{cf}[(TiO_2)_n] = \frac{E[(TiO_2)_n]}{n} - E[(TiO_2)]$$

where $E[(TiO_2)n]$ and $E[(TiO_2)]$ are the total energies including ZPE corrections of (TiO_2)n cluster and the free TiO_2 molecule respectively. TDDFT/B3LYP calculations are performed to evaluate vertical excitation energies to the lowest singlet and triplet excited states. The vertical ionization potentials (IPv) and vertical electron affinity (EAv) are calculated by single point B3LYP/6-31+G(d) calculations at optimized neutral cluster geometries for singly positively charged and negatively charged clusters.

$$IP_v = E[(TiO_2)_n^+] - E[(TiO_2)_n]$$

$$EA_V = E[(TiO_2)_n] - E[(TiO_2)_n^-]$$

Here $E[(TiO_2)_n^+]$, $E[(TiO_2)_n^-]$ and $E[(TiO_2)_n]$ are the total energies of positively charged, negatively charged and neutral clusters respectively without zero point correction.

6.3. Results and Discussion

A. Structures of (TiO_2)n clusters

It must be emphasized that the SA/GA level of calculation is done using the empirical potential already discussed. This potential indeed supports a linear structure for an isolated TiO_2 system. However, when supplying the initial guess geometry for DFT level of calculation, we supply the coordinates obtained from SA/GA as it is, as well as a second calculation, in which the SA/GA geometry is slightly perturbed. This is done so as to keep provision for the system to come out of a very high symmetric structure (like the linear) if it is needed for the DFT level of calculation. In most of the cases the SA/GA geometry, as it is or the perturbed one gives the same result but for some specific cases like the single TiO_2 unit the perturbed geometry helps the DFT calculation to recover the lower symmetry C_{2v} structure.

Table 6.2. Geometry and Properties of TiO_2

properties	Experimental[a]	This work	Calatayud[a]	Qu and Kroes[b]
d (Ti—O)Å	1.62	1.64	1.64	1.658
∠Ti-O-Ti°	110	111.8	110.9	110.8
ν (Ti—O)cm^{-1}	965	1033	1037	1026
IPa eV	9.54	9.68	9.86	9.75
EAa eV	1.59	1.68	1.58	1.69
H—L gap eV	2.2[c]	3.81	3.88	
S—T gap eV	1.96	2.05	2.01	

Compared to experimental and theoretical results from Refs. [56], [47], and [61].

[a] Ref. [56]. [b] Ref. [47]. [c] Ref. [61]

The monomer of TiO_2 has a bent C_{2v} structure. Its computed Ti=O bond length and O=Ti=O bond angle are 1.64Å and 111.8^0 respectively. The geometry and properties of

TiO_2 is compared with other workers in Table 6.2. The global minimum of dimer possessing C_{2h} symmetry is a non coplanar four membered ring with two bridging oxygen and two terminal oxygens having trans conformation is in agreement with previous calculations[6,42, 43, 45, 47, 49, 50,54, 56, 66]. The most stable structure of the trimer is shown in Fig.6.1. which has Cs symmetry. Inspection of the structure clearly shows that the geometry has one 3-fold coordinated Ti (Ti_{3c}) and two 4-fold coordinated Ti (Ti_{4c}). Tang et al. [60], Li-Dixon[6], Zhang[65] et al. found the triangular C_S structure as the global minimum. The global minimum for the tetramer corresponds to a C_{2v} structure with six 2-fold coordinated oxygens and four 4-fold coordinated Ti atoms. The terminal oxygens are oriented in cis fashion. This C_{2v} structure was predicted to be the lowest energy structure by Li & Dixon[6], Calatayud et al.[64] and Tang[68]. But Qu and Kroes[47] reported another C_{2v} configuration with only one and two coordinated oxygen atoms as the global minimum. Marom et al.[103] employed basin hopping and DFT to investigate structure and electronic properties of TiO_2 clusters. The global minimum predicted by our SA/GA+DFT method agrees with that of Marom et al.[103]. The lowest energy structures obtained from SA/GA + DFT for n=1-4 are those obtained by Zhao et al.[104]. The global minimum structure of the pentamer has Cs symmetry where all the titanium atoms are 4-fold coordinated and considering O atoms, seven are bridging, two are terminal and the last one oxygen is 4-fold coordinated. It is the most stable isomer in the work of Qu & Kroes[47], Calatayud et al.[56], Tang et al[60]. The lowest energy motif of $(TiO_2)_6$ is of C_2 symmetry. The geometry is characterised by six 4 fold coordinated Ti atoms and one 4 fold coordinated O atom with two terminal oxygens. This structure is reported as global minimum by Qu[47], Calatayud[56] and Tang et al.[60]. The global minimum of $(TiO_2)_7$ is of C_{2v} symmetry involving 10 bridging oxygens, two 3 fold coordinated oxygen and two terminal oxygen. It is the lowest energy structure reported by Qu[47] and Calatayud[56]. However, Shevlin and Woodley [55] reported a structure of C_2

symmetry as the global minimum which is 16.1kcal/mole higher in energy than the C_{2v} structure obtained by our SA/GA + DFT study. For $(TiO_2)_8$ the global minimum is of C_2 symmetry involving eight 4-fold coordinated Ti atom, twelve bridging oxygens, two 3-fold coordinated oxygens and two terminal oxygens. It is reported as the lowest energy structure by Syzgantseva et al.[61]. However, Zhang et al.[65] and Qu[47] reported a structure with one terminal oxygen as the lowest energy structure. Hamad et al. claimed a compact structure of C_2 symmetry as the global minimum which is 69.5kcal/mole higher in energy than the global minimum obtained by SA/GA + DFT study. The nonamer i.e; $(TiO_2)_9$ shows an unsymmetrical structure of C_1 point group. The structure consists of 36 bonds with six 4-fold Ti atom and three 5-fold Ti atom. The geometry is characterised by two oxygen having 3-fold coordination, two perpendicular terminal oxygen, and 14 bridging oxygen atoms. This is the global minimum suggested by Hamad et al.[46]. However Qu and Kroes[47] claims a Cs structure as the global minimum. The lowest energy conformer of the decamer $(TiO_2)_{10}$ shows an unsymmetric structure containing two 3-fold coordinated oxygen and one terminal oxygen. To the best of our knowledge, this structure is not so far reported in literature. However Shevlin et al.[55] reported a C_{2v} structure as the global minimum which is 12.3kcal/mole higher in energy than the global minimum obtained by SA/GA+DFT study as shown in Fig. 6.2. Again our global minimum structure is 13.1 kcal/mole more stable than the symmetric C_2 structure obtained by Qu[48] as the lowest energy conformer. The hendecamer i.e; $(TiO_2)_{11}$ has an unsymmetrical structure of C_1 point group involving one terminal oxygen. This structure is not reported in literature, again to the best of our knowledge. The global minimum of dodecamer $(TiO_2)_{12}$ is of D_{2d} symmetry involving 12 four-fold coordinated Ti atom and 24 bridging oxygen atoms. Qu and Kroes[48] found this symmetric D_{2d} structure as the global minimum.

N	Global and local minimum structures of (TiO_2)n Clusters
1	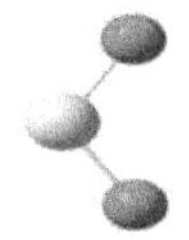 0.00
2	 0.00 16.6
3	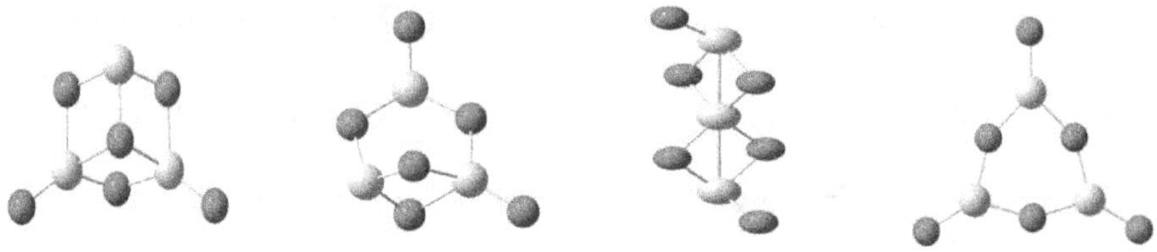 0.00 8.2 9.9 22.5
4	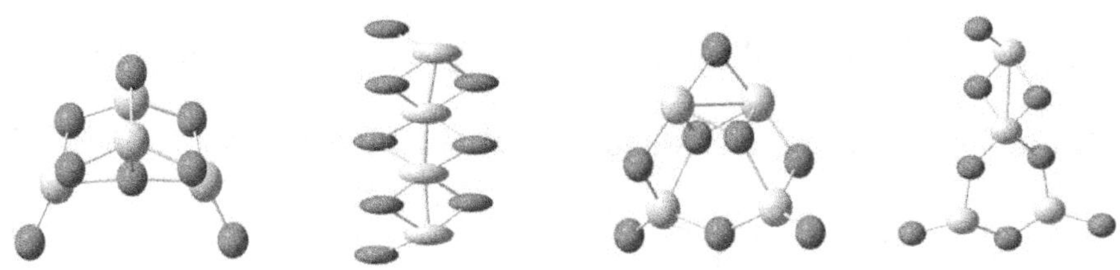 0.00 21.8 33.4 34.3
5	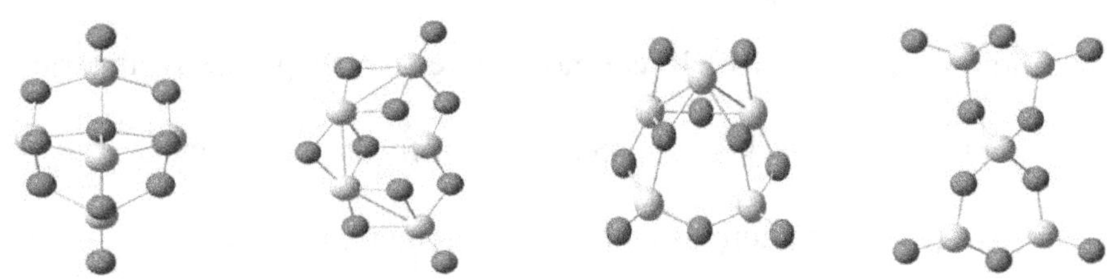 0.00 0.4 45.6 56.3
6	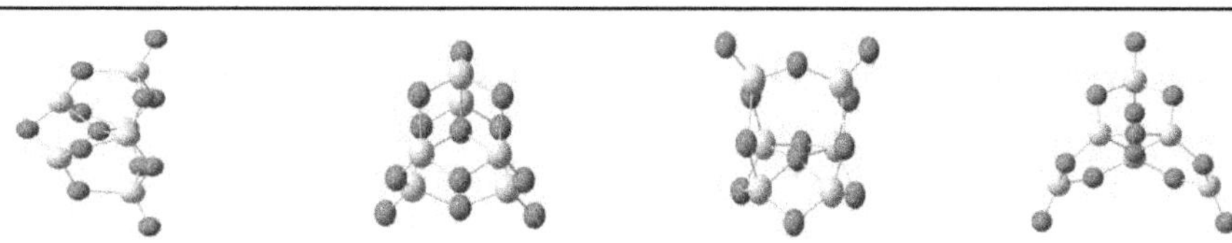 0.00 14.1 18.3 28.9

Fig.6.1 Structures of (TiO_2)n Clusters with n=1-6

N	Global and local minimum structures of $(TiO_2)n$ Clusters
7	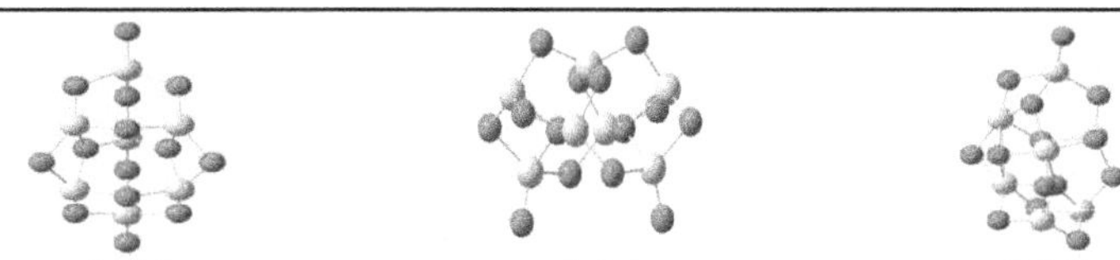 0.00 16.1 24.7
8	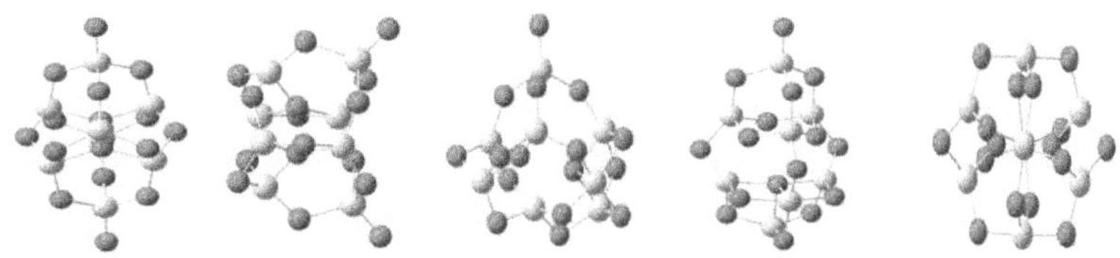 0.0 5.8 32.9 37.0 69.5
9	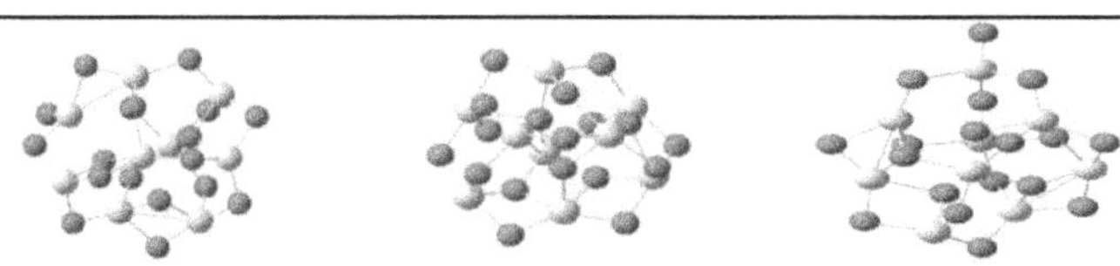 0.0 10.9 12.4
10	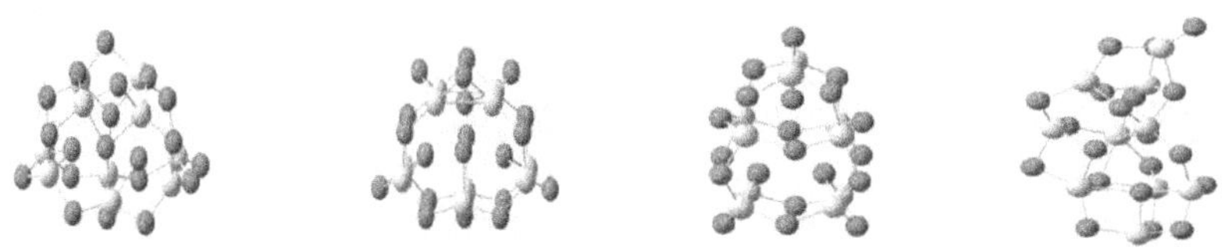 0.0 12.3 13.1 29.6
11	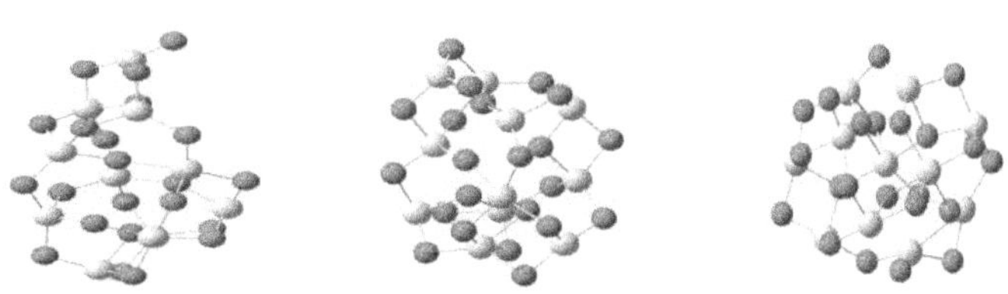 0.0 5.5 23.6
12	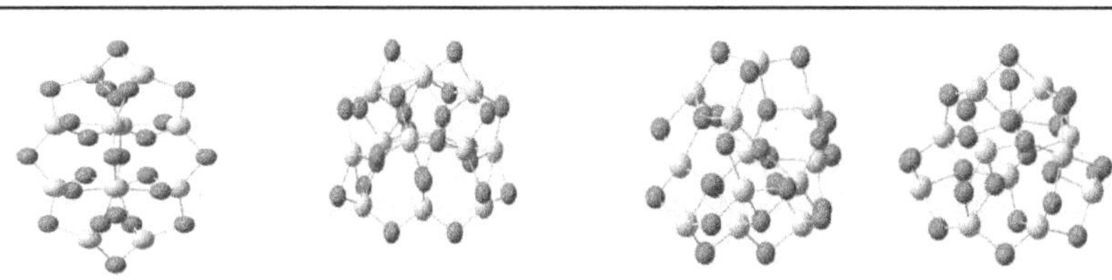 0.0 16.1 17.1 29.6

Fig.6.2 Structures of $(TiO_2)n$ Clusters with n=7-12

For all clusters at least 4 isomers from SA-GA are optimized through DFT calculation. The energies of global ones are taken as 0.00kcal/mole and the energies of local structures are reported relative to global ones as shown in Fig. 6.1 and Fig. 6.2 .We in our work while doing the quantum chemical calculation subjected not only the global minimum but also some of the low lying local minima after the GA stage to quantum chemical calculation. While in some sizes there is complete agreement but in some (n=6, 7, 10, 11 and 12) we were able to obtain slightly lower energy structure after the quantum chemical calculation. This is due to the use of global and some of the local minimum at the quantum chemical calculation stage.

B. Electronic Properties of (TiO_2)n clusters

For TiO_2 nanoparticles, we have reported Cluster formation energies (Ecf), HOMO-LUMO gap (Eg=[LUMO-HOMO]), vertical ionization potential (IPv), vertical electron affinity (EAv) and vertical excitation energy (Eve) in Table 6.3. The variation of these quantities with cluster size is displayed graphically in Fig. 6.3. It is clearly seen from the figure that cluster formation energy increases in magnitude initially for the smaller clusters and for the larger ones it shows constancy as expected. The HOMO-LUMO gap shows odd-even oscillation with the change of cluster size. The frequency of oscillation decreases with the increase of cluster size which is clearly visible in Fig. 6.3. An interesting odd-even oscillation of the vertical ionization potential of (TiO_2)n clusters is noticed in the Fig 6.3. The IPv decreases with increase of cluster size as well as the frequency of oscillation decreases. This suggests that removal of an electron is easier for clusters of larger size. The vertical excitation energy i.e; energy required in going from ground singlet state to first excited singlet shows similar trend, i.e; rate of fluctuation diminishes slowly with the increase of cluster size. The computed vertical electron affinity (EAv) shows an increasing trend with the increase of cluster size. From the structural details of clusters we can comment that even-n clusters tend to form compact structures of

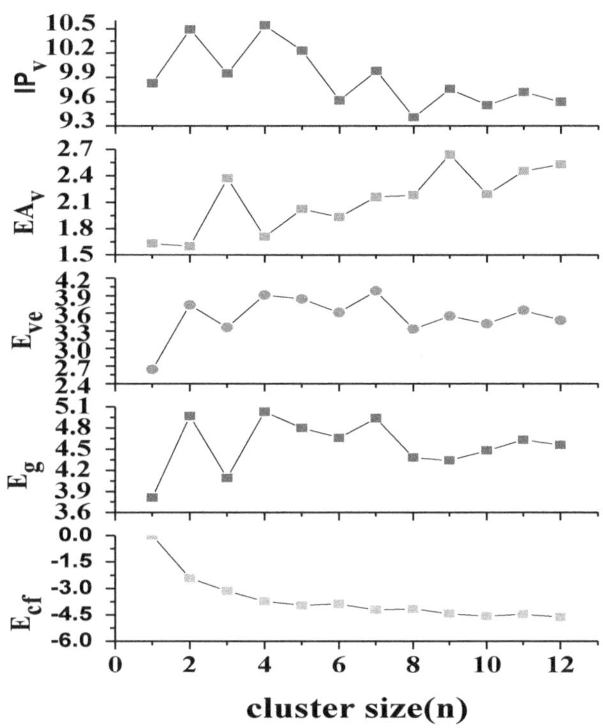

Fig. 6.3 variation of electronic properties with size(n)

more covalent nature that are more stable while odd-n clusters tend to form compact structures of more ionic nature that are less stable. The general trend of the decrease of HOMO-LUMO gap (Eg=[LUMO-HOMO]), Singlet-Triplet gaps(Est), Vertical Excitation Energies (Eve) with the increase of cluster size may be probably due to greater number of bridging oxygen atoms in association with larger number of titanium atoms in clusters of higher size. The clusters showing large HOMO-LUMO gap indicates a stable compound while those with lower gap values show the sign of greater reactivity [56]. Small sized $(TiO_2)_n$ clusters can be analyzed from the point of their stability and hence propensity to react with external agents (acids like H^+ or base like NH_3)[1, 3, 56, 62] from the magnitude of its HOMO-LUMO gap. It is observed that the HOMO-LUMO gap for small clusters do not show a monotonic change but an oscillating feature with change of size. Structurally if a (TiO_2)n cluster has under coordinated O or Ti atoms (than what is maximum possible) the HOMO goes up and the LUMO comes down in energy, hence reducing the HOMO-LUMO gap[56,62]. A larger extent of bonding/coordination has just the opposite effect. In a situation like this the HOMO becomes principally O like in character with lone pair of electrons which it can use with ease while involved in chemical combination with an electron deficient species. So the HOMO in a situation like this is the basic centre of the (TiO_2)n cluster. On the other hand the LUMO in these low HOMO-LUMO gap systems is predominantly Ti in character with electron deficiency. This centre serves as the acidic site and can coordinate with a electron rich external species.

Table 6.3. B3LYP/6-31+G(d)-Calculated cluster formation energies per TiO_2 unit (with ZPE correction), HOMO-LUMO Gaps (Eg), vertical ionization potential(IPv) and TDDFT/ B3LYP Vertical Excitation Energies(E_{ve}), singlet-triplet gaps (E_{st}) for the global minimum of (TiO_2)n clusters in eV units, bulk rutile values are given in last row(a-e)

N	E_{cf}	E_g	Eg(Qu) [a]	E_{ve}	E_{st}	EA_v	EA_v(Qu) [b]	IP_v	IP_v(Qu) [a]
1	0.00	3.81		2.64	2.05	1.63	1.63	9.83	
2	-2.43	4.97	4.89	3.74	3.27	1.60	1.60	10.49	10.44
3	-3.15	4.09	3.69	3.36	1.97	2.37	2.66	9.95	10.13
4	-3.73	5.03	3.15	3.91	2.84	1.71	1.96	10.54	9.35
5	-3.96	4.80	4.54	3.84	2.13	2.02	2.03	10.23	10.44
6	-3.89	4.66	4.53	3.61	1.87	1.93	2.60	9.62	10.49
7	-4.21	4.94	3.96	3.98	2.82	2.16	2.46	9.98	10.19
8	-4.18	4.38	3.58	3.33	2.05	2.18	2.59	9.41	9.50
9	-4.43	4.34	3.84	3.55	2.29	2.64	2.69	9.76	9.94
10	-4.58	4.48	4.15	3.42	2.53	2.19	2.66	9.56	9.52
11	-4.47	4.63	4.24	3.65	2.42	2.45		9.72	9.56
12	-4.63	4.56	4.65	3.48	2.89	2.53		9.60	9.98
Buk	6.62[c]	3.03[c]		3.0[c]	3.05[e]				

[a] Ref. 47 [b] Ref. 106 [c] Ref. 48 [d] Ref. 105 [e] Ref. 55

So we can come to a reasonably good conclusion that the sizes of (TiO_2)n, which have low HOMO-LUMO gap are less stable and show stronger affinity for reacting than the ones with higher HOMO-LUMO gap[56,57,62]. The HOMO-LUMO gap is observed to approach the bulk limit at n=6 and shows constancy for n=6-10 within the experimental limit [36, 53]. In case of $(TiO_2)_n$ clusters the HOMO is distributed mainly about oxygen anions and the LUMO about titanium atoms. The HOMO has main contribution from O-2p orbital while Ti-3d orbital contributes predominantly to the LUMO [107]. The bonding interaction between titanium and oxygen in $(TiO_2)_n$ clusters is strong with both ionic and covalent characters[48]. Qualitatively in $(TiO_2)_n$ clusters the singlet-triplet gap originates from electron transfer from oxygen 2p orbital to titanium 3d orbital [64].

C. IR spectroscopic features of (TiO_2)n clusters

After finding the structures, we have gone ahead to evaluate the infrared (IR) spectroscopic peaks of the $(TiO_2)_n$ cluster systems, with n=1-12. Absence of imaginary frequencies indicates that the structures are minimum on the potential energy surface. Our calculation of IR peaks is based on the normal mode analysis carried out with the global structure of each of the specific sizes. Broadly, the peaks shown by any TiO_2 cluster system can be classified into three categories. Very intense lines in the region of 980-1020cm^{-1} signifies the presence of terminal Ti-O bonds, while the stretch due to Ti-O-Ti bridging group occurs at below 930cm^{-1} and if threefold coordinated oxygen atoms are present, peaks can appear at around 530-760cm^{-1}. The infrared (IR) spectrums of all the clusters are shown in Fig. 6.4(a) to (l). The line width used is the default in Gauss View 5.0 package. The bare TiO_2 unit obviously has peaks at around 1000cm^{-1} and below 930cm^{-1} are seen which signifies of both bare Ti-O stretch and the existence of Ti-O-Ti bridging group. The same is the situation for $(TiO_2)_3$ clusters. The structures at the calculation SA/GA stage as well as after the DFT calculations confirm the spectral predictions in accordance with the geometry obtained. From clusters of size $(TiO_2)_4$ and higher, peaks can be seen in a very broad range of wave numbers, right from 500cm^{-1} to upto 1000cm^{-1}. This is because from $(TiO_2)_4$ one observes the presence of all the three major bonding traits, i.e; terminal Ti-O, bridging Ti-O-Ti and the presence of 3 or 4 coordinated oxygen atoms. Another feature which is worth mentioning is the fact that the intensity of the peak due to terminal Ti—O gets diminished in intensity at the expense of the Ti-O-Ti or the 3 or 4 coordinated oxygen's for cluster of sizes n=9 and higher.

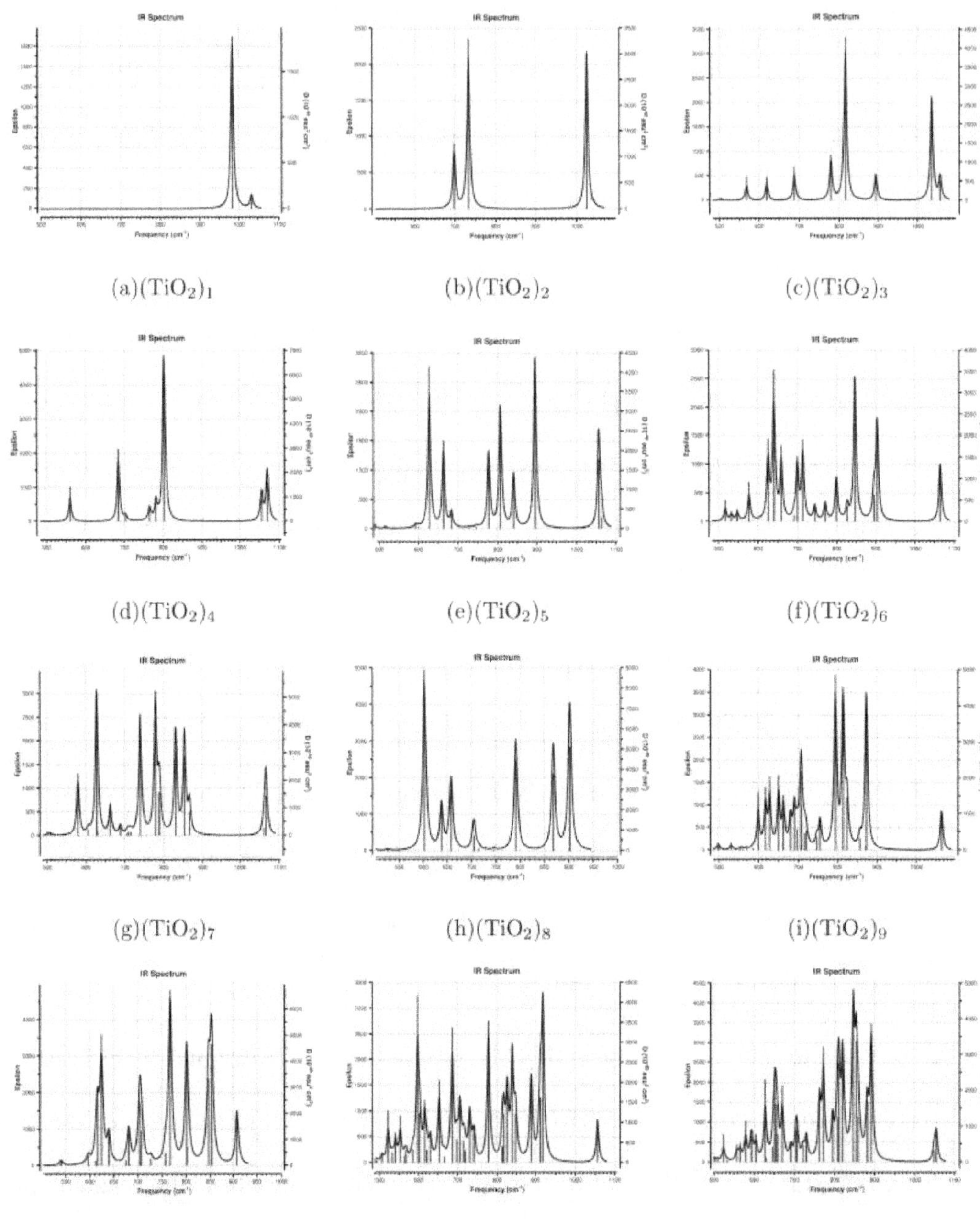

(a)$(TiO_2)_1$ (b)$(TiO_2)_2$ (c)$(TiO_2)_3$

(d)$(TiO_2)_4$ (e)$(TiO_2)_5$ (f)$(TiO_2)_6$

(g)$(TiO_2)_7$ (h)$(TiO_2)_8$ (i)$(TiO_2)_9$

(j)$(TiO_2)_{10}$ (k)$(TiO_2)_{11}$ (l)$(TiO_2)_{12}$

Fig. 6.4. Infrared (IR) spectra of (TiO_2)n clusters

This is understandable as with the increase in the size of the cluster the number of Ti-O-Ti units as well as the 3 or 4 coordinated oxygen centres increases at the expense of the terminal Ti—O units. Bulk rutile shows absorptions around 357, 418, 539 and 652cm^{-1}[48], whereas for anatase it is around 360, 540, 660 and 740cm^{-1}[48].

6.4 Conclusion

Using a stochastic search involving (genetic algorithm / simulated annealing) followed by density functional calculations we have identified lowest energy structures for small nanoclusters of titanium oxide. We see that with the increase in size titania clusters relax so that titanium atoms are in a more bulk like environment with the consequent formation of dangling oxygen atoms. For increasing cluster size (n), compact structures with one or two terminal O-atoms become increasingly energetically favoured. The global minimum structures show infrared (IR) peaks at 980-1020cm^{-1} due to terminal Ti-O bonds and below 930cm−1 due to Ti—O—Ti bridging bonds may be observed, with peaks in the range530-760cm^{-1} due to 3-fold coordinated oxygen atoms. The even-n (TiO_2)n cluster tend to form compact covalent networks consisting of 4-coordinated $Ti^{(4)}$ and 2-coordinated $O^{(2)}$ atoms while odd-n clusters tend to form more ionic structures with highly coordinated Ti and O atoms[48]. In consistence with the odd-even oscillation of structural features of (TiO_2)n clusters, various electronic properties like HOMO-LUMO gaps, vertical excitation energies , ionization potentials also show odd-even oscillations. The cluster formation energy increases resulting in higher stability as the cluster size grows. The odd-n (TiO_2)n cluster exhibit small vertical excitation energies which in some cases becomes significantly smaller than the band gap of bulk TiO_2. Utilizing this materials can be engineered to incorporate (TiO_2)n nanoparticles which could exhibit visible light photo activity[49].

6.5 References

[1] S. Li, H. J. Zhai, L. S. Wang, D. A. Dixon, J. Phys. Chem. A **2012**, 116, 5256

[2] S. Li, J. M. Hennigan, D. A. Dixon, J. Phys. Chem. A **2009**, 113, 7861

[3] S. Li, D. A. Dixon, J. Phys. Chem. A **2006**, 110, 6231

[4] S. Li, D. A. Dixon, J. Phys. Chem. A **2007**, 111, 11093

[5] S. Li, D. A. Dixon, J. Phys. Chem. A **2007**, 110, 11908

[6] S. Li, D. A. Dixon, J. Phys. Chem. A **2008**, 112, 6646

[7] H. J. Zhai, B. Kiran, L. F. Cui, X. Li, J. Li, L. S. Wang, J. Am. Chem. Soc **2004**, 126, 16134

[8] H. J. Zhai, S. Li, D. A. Dixon, L. S. Wang, J. Am. Chem. Soc **2008**, 130, 5167

[9] S. Li, H. J. Zhai, L. S. Wang, D. A. Dixon, J. Phys. Chem. A **2009**, 113, 11273

[10] S. Li, D. A. Dixon, J. Phys. Chem. C **2011**, 115, 19190

[11] S. Li, L. C. Guenther, M. S. Kelley, D. A. Dixon, J. Phys. Chem. C **2011**, 115, 8072

[12] H. J. Zhai, B. Wang, X. Huang, L. S. Wang, J. Phys. Chem. A **2009**, 113, 3866

[13] H. J. Zhai, L. S. Wang, Chem. Phys. Lett **2010**, 500, 185

[14] M. Gratzel, Nature **2001**, 414, 338

[15] A. Fujishima, K. Honda, Nature **1972**, 238, 37

[16] A. L. Linsebigler, G. Lu, J. T. Yates, Chem. Rev. **1995**, 95, 735

[17] U. Dicbold, Surf. Sci. Rep. **2003**, 48, 53

[18] A. G. Agrios, P. Pichat, J. Appl. Electrochem. **2005**, 35, 655

[19] M. R. Hoffmann, S. T. Martin, W. Choi, D. Bahnemann, Chem. Rev. **1995**, 95, 69

[20] A. M. Volodin, Catal. Today. **2000**, 58, 103

[21] T. L. Thompson, J. T. Yates, Jr. Chem. Rev **2006**, 106, 4428

[22] A. Hagfeldt, M. Gratel, Chem. Rev **1995**, 95, 49

[23] Y. Wang, N. Herron J. Phys. Chem **1991**, 95, 525

[24] V. M. Aroutiounian, V. M. Arakelyan, G. E. Shahnazaryan, Sol. Energy. **2005**, 78, 581

[25] H. Liire, K. Sunanda, K. Hashimoto, Electrochemistry **2004**, 72, 807

[26] D. Bahnemann, Sol. Energy **2004**, 77, 445

[27]A. Vohra, D. Y. Goswami, D. A. Deshpande, S. S. Block J. Ind. Microbiol. Biotechnol **2005**, 32, 364

[28] P. Fernandez, J. Blanco, C. Sichel, S. Malato, Catal. Today **2005**, 101, 345

[29] V. A. Schwarz, S. D. Klein, R. Hornung, R. Knochenmuss, P. Wyss, D. Fink, U. Haller, H. Watt, Lasers. Surg. Med. **2001**, 29, 252

[30] Z. Fang, D. A. Dixon, J. Phys. Chem. A **2013**, 117, 3539

[31] N. S. Mclntyre, K. R. Thompson, W. Weltner, J. Chem. Phys **1971**, 75, 3243

[32] G. Balducci, G. Gigli, M. Guido, J. Chem. Phys. **1985**, 83, 1909

[33] W. Yu, R. B. Freas, J. Am. Chem. Soc. **1990**, 112, 7126

[34] B. C. Guo, K. P. Kerns, A. W. Castleman, Int. J. Mass. Spectrom. Ion. Processes. **1992**, 117, 129

[35] G. V. Cherinthin, I. Andrews, J. Phys. Chem. **1995**, 99, 6356

[36] H. Wu, L. S. Wang, J. Chem. Phys. **1997**, 107, 8221

[37] M. Foltin, G. J. Stueber, E. R. Bernstein, J. Chem. Phys. **1999**, 111, 9577

[38] Y. Matsuda, E. R. Bernstein, J. Phys. Chem. A. **2005**, 109, 314

[39] T.-H. Wang, L. Navarrete, M. Alejandara, S. Li, D. A. Dixon, J. L. Gole, J. Phys. Chem. A. **2010**, 114, 7561

[40] M. V. Ramana, D. H. Philips, J. Chem. Phys. **1988**, 88, 2637

[41] R. Bergstrom, S. Lunell, L. A. Eriksson, Int. J. Quant. Chem. **1996**, 59, 427

[42] M. B. Walsh, R. A. King, H. F. Schaefer III, J. Chem. Phys. **1999**, 110, 5224

[43] T. Albaret, F. Finocchi, C. Noguera, J. Chem. Phys. **2000**, 113, 2238

[44] T. Albaret, F. Finocchi, C. Noguera, Faraday Discuss **1999**, 114, 285

[45] S. M. Woodley, S. Hamad, J. A. Mejias, C. R. A. Catlow, J. Mater. Chem. **2006**, 16, 1927

[46] S. Hamad, C. R. A. Catlow, S. M. Woodley, S. Lago, J. A. Mejas, J. Phys. Chem. B. **2005**, 109, 15741

[47] Z.-W. Qu, G.-J. Kroes, J. Phys. Chem. B **2006**, 110, 8998

[48] Z.-W. Qu, G.-J. Kroes, J. Phys. Chem. C **2007**, 111, 16808

[49] K. S. Jeong, C. Chang, E. Sedimayr, D. Sulzle, J. Phys. B. At. Mol. Opt. Phys. **2000**, 33, 3417

[50] A. Hagfeldt, R. Bergstrom, H. O. G. Siegbahn, S. Lunell, J. Chem. Phys **1993**, 97, 12725

[51] T.-H. Wang, Z. Fang, N. W. Gist, S. Li, D. A. Dixon, J. L. Gole, J. Phys. Chem. C **2011**, 115, 9344

[52] A. C. Tsipis, C. A. Tsipis Phys. Chem. Chem. Phys. **1999**, 1, 4453

[53] H. J. Zhai, L. S. Wang, J. Am. Chem. Soc **2007**, 129, 3022

[54] X. Wei, R. Skomski, B. Balamurugan, Z. G. Sun, S. Ducharme, D. J. Sellmyer, J. Appl. Phys. **2009**, 105, 07C517

[55] S. A. Shevlin, S. M. Woodley, J. Phys. Chem. C. **2010**, 114, 17333

[56] M. Calatayud, L. Maldonado, C. Minot, J. Phys. Chem. C. **2008**, 112, 16087

[57] M. Calatayud, C. Minot, J. Phys. Chem. C **2009**, 113, 12186

[58] I. Bandyopadhyay, C. M. Aikens J. Phys. Chem. A **2011**, 115, 868

[59] A. S. Barnard, S. Erdin, Y. Lin, P. Zapol, J. W. Halley Phys. Rev. B **2006**, 73, 205405

[60] L. Tang, L. Sai, J. Zhao, R. Qiu, J. Comput. Chem **2012**, 33, 63

[61] O. A. Syzgantseva, P. Gonzalez-Navarrete, M. Calatayud, S. Bromley, C. Minot J. Phys. Chem. C **2011**, 115, 15890

[62] D. Cakir, O. Gulseren, J. Phys. Condens. Matter **2012**, 24, 305301

[63] L. Chiodo, M. Salazar, A. H. Romero, S. Laricchia, F. D. Sala, A. Rubio, J. Chem. Phys **2011**, 135, 244704-1

[64] T. Tsuchiya, J. L. Whitten, J. Chem. Phys. **2011**, 134, 114701

[65] W. Zhang, Ye. Han, S. Yao, H. Sun, Mat. Chem. Phys. **2011**, 130, 196

[66] T. Albaret, F. Finocchi, C. Nogurea, Appl. Surf. Sci **1999**, 144, 672

[67] D. J. Mowbray, J. I. Martinez, J. M. Garca-Lastra, K. S. Thygesen, K. W. Jacobsen, J. Phys. Chem. C **2009**, 113, 12301

[68] T. H. Rana, P. Kumar, A. K. Solanki, R. Skomski, A. Kashyap, J. Appl. Phys. **2013**, 113, 17B526

[69] S. Kirkpatric, C. D. Gelatt, M. P. Vecchi, Science **1983**, 220, 671

[70] S. Kirkpatric, J. Stat. Phys. **1984**, 34, 975

[71] D. E. Goldberg, Genetic Algorithm in Search, Optimization and Machine learning; Addison Wesley; Reading, MA, **1989**

[72] J. Holland, Adaptation in Natural and Artificial Systems; MIT Press; Cambridge, MA, **1998**

[73] M. Mitchell, An introduction to Genetic Algorithm; MIT Press: Cambridge, MA, **1998**

[74] D. M. Deaven, K. M. Ho, Phys. Rev. Lett. **1995**, 75, 288

[75] D. M. Deaven, J. R. Morris, K. M. Ho, N. Tit, Chem. Phys. Lett **1996**, 256, 195

[76] W. J. Pullan, J. Chem. Inf. Modelling **1997**, 37, 1189

[77] W. J. Pullan, Comp. Phys. Comm. **1997**, 107, 137

[78] W. J. Pullan, J. Comput. Chem. **1997**, 18, 1096

[79] J. A. Niesse, H. R. Mayne, Chem. Phys. Lett. **1996**, 261, 576

[80] J. A. Niesse, H. R. Mayne, J. Chem. Phys. **1996**, 105, 4700

[81] J. A. Niesse, H. R. Mayne, J. Comput. Chem **1997**, 18, 1233

[82] R. L. Johnston, Dalton. Trans. **2003**, 4193

[83] C. Roberts, R. L. Johnston, N. T. Wilson, Theo. Chem. Acc. **2000**, 104, 123

[84] R. Ismail, R. L. Johnston, Phys. Chem. Chem. Phys. **2010**, 12, 8607

[85] K. Michaelian, Chem. Phys. Lett **1998**, 293, 202

[86] I. L. Garzon, N. Rendon, K. Michaelian, Phys. Rev. B. **1999**, 60, 2000

[87] K. Michaelian, Am. J. Phys. **1998**, 66, 231

[88] S. Ganguly Neogi, P. Chaudhury, J. Comput. Chem. **2012**, 33, 629

[89] S. Ganguly Neogi, P. Chaudhury, J. Comput. Chem. **2013**, 34, 471

[90] B. Hartke, Chem. Phys. Lett. **1995**, 240, 560

[91] B. Hartke, Chem. Phys. Lett. **1996**, 258, 144

[92] S. K. Gregurick, M. H. Alexander, B. Hartke, J. Chem. Phys. **1996**, 104, 2684

[93] B. Hartke, M. Schutz, H. J. Werner, Chem. Phys **1998**, 239, 561

[94] M. Chen, D. A. Dixon, J. Chem. Theo. Comput **2013**, 9, 3189

[95] Gaussian 09, Revision B.01, M. J. Frisch, Gaussian, Inc., Wallingford CT, **2010**.

[96] M. Matsui, M. Akaogi, Mol. Simul. **1991**, 6, 239

[97] A. D. Becke, J. Chem. Phys. **1993**, 98, 5648

[98] C. Lee, W. Yang, R. G. Parr, Phys. Rev. B **1988**, 37, 785

[99] T. Clark, J. Chandrasekhar, G. W. Spitznagel, P. v. R. Schyeler, J. Comput. Chem **1983**, 4, 294

[100] V. A. Rassolov, J. A. Pople, M. A. Ratner, T. L. Windus, J. Chem. Phys. **1998**, 109, 1223

[101] A. J. H. Wachters, J. Chem. Phys. **1970**, 52, 1033

[102] P. J. Hay, J. Chem. Phys. **1977**, 66, 4377

[103] N. Marom, M. Kim, J. R. Chelikowsky, Phys. Rev. Lett. **2012**, 108, 10608-1

[104] H. M. Zhao, M. M. Wu, Q. Wang, P. Jena, Physica B **2011**, 406, 4322

[105] H. Tang, F. Levy, H. Berger, P. E. Schmid Phys. Rev. B **1995**, 52, 7771

[106] Z. –W. Qu, G.-J. Kroes, J. Comput. Chem. **2010**, 31, 2038

[107] S. K. Sahoo, S. Pal, P. Sarkar, C. Majumder, Chem. Phys. Lett **2011**, 516, 68

Chapter 7

Concluding Remarks

As concluding remarks we would like to summarize about the objectives achieved while solving the problems discussed in the thesis and also try to discuss the logical extensions that one can pursue in the future.

We in the present study have through out explored the idea of, combining stochastic optimization based search techniques and the impressive quantum chemical methods that are already at our disposal. To put it in the right perspective we have solved all the problems using a two tier strategy. The first has always being to use good quality empirical potentials to describe the chemical systems of interest which have all been clusters of different types. This first step has been taken purposely because an initial exploration on the empirical potential energy surface can very quickly give us good quality structures which can later on be used as inputs for doing the quantum chemical calculation. These empirical potential functions, though apparently innocuous, can just be the opposite when one gets down working with them. Now, the use of stochastic optimization is crucial because stochastic optimizers are better suited to find out global and the low lying local minima as opposed to deterministic ones, for reasons which have been already discussed in the thesis. Throughout our study we have seen that the structures supported by the empirical potential, if used as starting candidate structures in a quantum chemical calculation, the convergence is extremely quick with the additional important guarantee of achievement of global minima. This two tire strategy in our opinion reduces computational cost and guarantees achievement of good solution in the end.

If one seriously wants to come to any conclusion about the effectiveness of any proposed strategy, one must put it to serious test under trying conditions. So, we decided to study various types of clusters with varying nature of the forces that binds the basic units. Chapters 2 and 3 is devoted to the study of clusters formed by hydrogen bonding, chapter 4 to formation of clusters essentially by dispersive forces (van der Waals) while chapter 5 studies exclusively ionic bonded units and chapter 6 is devoted to a cluster system in which both ionic and covalent forces play equally important roles. The fact that these varied situations have been equally well handed by our central strategy, instills in us a certain degree of confidence that our strategy has a certain amount of generality and can be a viable alternative to existing ones in the study of even more complicated systems in the future.

At this juncture while, we are coming to the end of this written document we cannot help but give our opinion about some future work which can be effectively handled with the proposed theoretical tool. If we look at the large amount of literature on empirical potentials, the common thread in all these is the presence of a Lenard-Jones term. Now unfortunately well calibrated Lenard- Jones potential parameter exists only for closed shell systems. So, the large variety of cluster systems which have one or more open shell units are outside the perview of standard Lennard-Jones potential. If one can design effective parameters for these systems and construct good quality empirical potentials, a new door will open for us to study cluster systems with quite different behaviour from the present ones. Last, but not the least, one must also carry on the quest for designing and using newer stochastic optimization tools as no single method can boast of being the panacea for solving all problems and being equally effective in all situations.

www.ingramcontent.com/pod-product-compliance
Lightning Source LLC
LaVergne TN
LVHW082342200926
848481LV00012B/3427